GROUP THEORY IN
SOLID-STATE PHYSICS

A. P. CRACKNELL

Carnegie Laboratory of Physics, University of Dundee, Scotland

TAYLOR & FRANCIS LTD
LONDON

HALSTED PRESS
division
John Wiley & Sons Inc.
NEW YORK—TORONTO

1975

First published 1975 **by** *Taylor & Francis Ltd., London and Halsted Press (a division of John Wiley & Sons Inc.), New York,*

Reprinted from ADVANCES IN PHYSICS, *Volume 23, No. 5, September 1974*

Taylor & Francis ISBN 0 85066 090 4

Printed and bound in Great Britain by Taylor & Francis (Printers) Ltd., Rankine Road, Basingstoke, Hants.

Library of Congress Cataloging in Publication Data

Cracknell, Arthur P.
Group theory in solid-state physics.
(Monographs on Physics)
Bibliography: p.
Includes indexes.
1. Solids. 2. Groups, Theory of. I. Title. II. Series.

QC176.C7   530.4'1   75–11959
ISBN 0–470–18320–9

Preface

This monograph, which was originally a review article, is intended to give the reader an appreciation of the current state of the various possible applications of group theory in solid-state physics. It is not a treatise and therefore those aspects of the subject which are well-established and adequately described already in the literature are not expounded in full detail, although of course their general principles are discussed (this philosophy is expounded in more detail in section 1.1.). The main intention is to show that, in addition to the well-established applications, there have now been several new developments in recent years. These new developments are described in some detail. The treatment is not restricted to the use of the classical point groups and space groups but, where magnetic ordering is important, the appropriate generalized symmetry groups are used.

I am grateful to many people with whom I have discussed various aspects of these applications of group theory, especially to Dr. C. J. Bradley, who being a mathematician has helped to "rigourize" many of my more intuitive ideas, and to Dr. W. G. Ferrier and Dr. A. G. Fitzgerald for specific help with the material of sections 6.3 and 7.4 respectively. I am also grateful to Professor B. R. Coles for his encouragement during the writing of the manuscript, to all the various authors, editors, and publishers who have given permission for the reproduction of copyright material, the sources of which are indicated *in situ*, to Mr. R. S. Paterson for drawing all the diagrams, and to Dr. J. A. Przystawa for some help in compiling the author index.

August, 1974.   A. P. CRACKNELL.

Contents

GROUP THEORY IN

SOLID STATE PHYSICS

1. Introduction

1.1. *A personal viewpoint*

During the past ten or fifteen years we have seen the publication of a number of books and review articles dealing with the description of the applications of group theory in solid-state physics. It would therefore seem to be appropriate to give some justification for offering to the public yet another article on this subject.

The foundations of the application of group theory in quantum mechanics were laid about 1930, principally by Weyl, Wigner and von Neumann (see Weyl (1931), the translation of the classic book by Wigner (1959) and the collected works of von Neumann (1961, 1963)), and group theory quickly came to be exploited in quantum-mechanical treatments of atomic and molecular physics. From that time, as the understanding of solid-state physics began to become more quantitative, the applications of group theory in the physics of crystalline solids began to be appreciated. From the early work on crystal-field theory by Bethe (1929) and on electronic band structures by Bouckaert *et al.* (1936), it became apparent that the key to much of the exploitation of symmetry in the quantum mechanics of a molecule or solid lay in the irreducible representations of the classical point groups and space groups. Tables of the irreducible representations of the point groups have been available for a long time and descriptions of their use in connection with the energy levels of molecules and the energy levels of ions in crystals have been given by a number of authors (Bethe 1929, Rosenthal and Murphy 1936, Eyring *et al.* 1944, Wilson *et al.* 1955, McClure 1959 a, b, Herzfeld and Meijer 1961, Judd 1963, Slater 1963, 1965, Hutchings 1964). A particularly convenient summary of the properties of the crystallographic point groups and their representations is given by Koster *et al.* (1963).

Probably the earliest detailed textbook on the applications of group theory in solid-state physics is the book by Bhagavantam and Venkatarayudu (1948), while many others have been published more recently (Nye 1957, Lomont 1959, Heine 1960, Jones 1960, Griffith 1961, Hamermesh 1962, Meijer and Bauer 1962, McWeeny 1963, Birss 1964, Knox and Gold 1964, Shubnikov and Belov 1964, Tinkham 1964, Koptsik 1966, Jansen and Boon 1967, Cracknell 1968 a, Cornwell 1969, Leech and Newman 1969, Boardman *et al.* 1973, Janssen 1973, Wooster 1973). In view of the undoubted importance of the irreducible representations of the space groups in various applications, a number of authors set out in the early 1960s to compile tables of these representations for all the 230 space groups (for references see § 5.1). Among these were Dr. C. J. Bradley and the present author who set out to write a book to serve three main purposes :

    (i) to set out in detail the mathematical theory of the determination and the properties of the irreducible representations of the (classical) point groups and space groups,

    (ii) to tabulate these irreducible representations for all the 32 classical point groups and the 230 classical space groups, and

    (iii) to describe the applications of this theory (i) and of these tables (ii) in solid-state physics.

As the writing of that book (Bradley and Cracknell 1972) progressed, other developments caused alterations in the plan of campaign. First, a considerable increase in the use of elastic neutron-scattering techniques has led to the extensive use of the Heesch–Shubnikov groups in describing the structures of numerous magnetically ordered crystals. It therefore became important to give a detailed consideration of the theory and properties of the irreducible co-representations of these non-unitary groups. Secondly, (iii) never materialized. This was partly because of the appearance of several books which dealt with these applications in considerable detail (see above). Moreover, whereas (i) and (ii) represented relatively closely circumscribed fields of knowledge, (iii) represented a more open-ended subject in which progress and developments were occurring quite quickly.

In the present article it is intended to remind the reader briefly of the general principles involved in the application of group theory to the physics of crystalline solids. We shall then go on to indicate in some detail the types of problem to which group theory can be applied and to describe the nature of the information which can be obtained from a group-theoretical approach to the problem. These applications include some cases in which the use of group theory is already extremely well-established and familiar to most solid-state physicists (these applications are described in § 2). In these cases no attempt will be made to give an exhaustive account because there would be no point in describing in detail, yet again, those things which are adequately described in so many excellent texts. A slightly different approach would have been to redistribute the material in § 2 between § 4.1 on crystal-field theory and § 6.1 on electronic band structures. This redistribution would obviously be done if one were to adopt a strictly formal approach to the subject. However, one of our aims is to consider the development or evolution which has occurred in the application of group theory in solid-state physics. It is our intention to show that the rôle of group theory in solid-state physics is changing so that, as well as mentioning the more established uses, we shall also describe some of those areas in which it has recently assumed a new importance, or looks as though it may soon become important. A good example of this evolution is provided by the case of electronic band structures. As the original use of group theory in this connection, namely in simplifying band-structure calculations, declines, it seems likely that a new use for group theory will be found in constructing parametrization schemes for Fermi surfaces (see §§ 6.1 and 6.2). Some other recently introduced examples of the use of group theory in connection with the energies of particles, or of quasiparticle excitations, in crystalline solids will be described in the later parts of § 6. Sections 7 and 8 will be devoted to the consideration of some recent developments involving the use of representation, or co-representation, theory for both non-magnetic and magnetically ordered materials. Certain parts of this article, in particular those sections which are concerned with magnetically ordered materials, describe applications which are considered in much greater detail in a book on magnetic crystals which is currently in the press (Cracknell 1974 a).

It may be useful to indicate the extent of the background knowledge of group theory which the reader of the present article is assumed to possess. It is not assumed that the reader has made an exhaustive study of the book on *The mathematical theory of symmetry in solids* by Bradley and Cracknell (1972). However, we shall assume that the reader has an acquaintance with the general outlines of some of the subject matter covered therein ; to be specific, we shall assume a familiarity with much of the subject matter of chapters 1 and 3, and some familiarity with the earlier parts of chapters 6 and 7.

It should perhaps be mentioned that as well as the move of group theory into new areas in solid-state physics, there is also now an increasing awareness of the relevance of topological concepts in various parts of solid-state physics ; in this connection we would refer the reader to the recent article by Killingbeck (1970).

1.2. *Classical symmetry and generalized symmetry*

Our knowledge of the structures of crystalline materials and of the various excited particle or quasiparticle states in these materials has mostly been obtained from various kinds of scattering experiments. These experiments may involve the scattering of beams of particles or the scattering of beams of photons, while in each case the scattering may be elastic or inelastic. The particles involved may be electrons or neutrons or, occasionally, protons, while the photons involved may be X-rays, optical, infra-red or ultra-violet photons. The elastic scattering, that is scattering without change of the energy (or wavelength) of the particle or photon, principally gives information about structures, while the inelastic scattering principally gives information about excited particle or quasiparticle states in the crystal. We shall take the term ' diffraction ' to be synonymous with elastic scattering and we shall postpone any discussion of inelastic scattering until § 7.

By far the most firmly established and most commonly used technique for structure determination is that of X-ray diffraction ; it has the advantage of being cheaper and easier to use than the elastic scattering of electrons or neutrons. X-ray diffraction has been used for over half a century in the determination of crystal structures and at the present time there are at least 9000 crystalline materials whose structures have been determined (for lists see Donnay et al. (1963), Nowacki et al. (1967), and Wyckoff (1963, 1964, 1965, 1966, 1968)) ; each of these structures has been assigned to one of the 230 ' classical ' or ' Fedorov ' space groups. Historically the possibility of the use of the scattering of beams of electrons or of neutrons by an array of atoms forming a solid has been appreciated for a long time, but for various technical reasons the application of these methods is much more recent. However, electron diffraction and neutron diffraction are important in certain special circumstances. Because of the magnetic interaction, neutron diffraction is used in the determination of the orientations of the magnetic moments of the atoms in magnetically ordered crystals, see, for example, Bacon (1962). Electron diffraction can, in principle, be used for this purpose too and this possibility has been demonstrated, with low-energy electrons, for antiferro-magnetic NiO (Palmberg et al. 1968, Palmberg et al. 1969, Suzuki et al. 1971, Hayakawa et al. 1971) ; however, this technique has not so far been widely used for magnetic structure determinations because for electrons the magnetic

Table 1. Some important characteristics for X-ray, neutron and electron scattering

Energy, E	Neutron (10 meV)	X-rays (10 keV)	Electrons (100 keV)	LEED (100 eV)
Wavelength, λ	1 Å	1 Å	0·05 Å	1 Å
Extinction length, d_{ext}	10^5 Å	10^4 Å	10^2–10^3 Å	5 Å
Absorption length, $1/\mu$	10^8 Å	10^5 Å	10^3–10^4 Å	10 Å
$1/\Delta\mu$	$> 10^8$ Å	30×10^5 Å	$3 \times (10^3$–$10^4)$ Å	10 Å

Note. $1/\Delta\mu$ is the absorption length for the case when a Bragg reflection is excited (anomalous transmission). (Adapted from Dederichs (1972).)

scattering is dominated by the Coulomb scattering. For electrons it is convenient to distinguish between low-energy electrons, such as those used in commercial low-energy electron diffraction (LEED) apparatus, with energies up to about 100 eV and 'high'-energy electrons, such as those used in electron microscopy, with energies up to about 100 keV. In order to help to put into perspective the distinctive aspects of X-ray, neutron and electron scattering we indicate in table 1 the order of magnitude of some of the most important quantities associated with the scattering of beams of particles or photons. For low-energy electrons the range of the electrons is very small (of the order of only a few atomic layers) and therefore low-energy electron diffraction is particularly important in the determination of the structural properties of surfaces.

The object of elastic-scattering experiments is to determine the positions of the atoms in the unit cell of a crystal. For any given space group the general equivalent positions, as well as the various sets of special equivalent positions, are listed in volume I of the *International tables for X-ray crystallography* (Henry and Lonsdale 1965). For example, for the space group *Pmmn* ($D_{2h}{}^{13}$) the general equivalent positions listed in the *International tables* are

$$x, y, z \; ; \quad \tfrac{1}{2}-x, y, z \; ; \quad x, \tfrac{1}{2}-y, z \; ; \quad \tfrac{1}{2}-x, \tfrac{1}{2}-y, z$$
$$\bar{x}, \bar{y}, \bar{z} \; ; \quad \tfrac{1}{2}+x, \bar{y}, \bar{z} \; ; \quad \bar{x}, \tfrac{1}{2}+y, \bar{z} \; ; \quad \tfrac{1}{2}+x, \tfrac{1}{2}+y, \bar{z}.$$

One can imagine each of these positions as being generated from the position x, y, z by some symmetry operation of the space group. A general symmetry operation of a crystal is a compound operation consisting of some point-group operation R followed by a displacement specified by a translation **v**. This means that an element of the space group of the crystal can be denoted by the symbol $\{R|\mathbf{v}\}$; the interpretation of this operation is that it acts on the points of space and that these are always referred to a fixed set of axes, that is it is an *active* operation. The effect of an operation denoted by $\{R|\mathbf{v}\}$ on a vector **r** can be represented by

$$\{R_1|\mathbf{v}_1\}\mathbf{r} = \mathbf{r}' = R_1\mathbf{r} + \mathbf{v}_1 \tag{1.1}$$

where $R_1\mathbf{r}$ is the vector that is produced from **r** by the application of the active point-group operator R_1. The symbols $\{R|\mathbf{v}\}$ were introduced by Seitz (1936) and they are called *Seitz space-group symbols*. For any given space group it is possible to use the coordinates of the equivalent general positions listed in volume I of the *International tables* to determine the symbols $\{R|\mathbf{v}\}$ for the operations of that space group as well as the physical nature of the symmetry element associated with each symbol $\{R|\mathbf{v}\}$. For example, for the space group *Pmmn* ($D_{2h}{}^{13}$) it is quite easy to identify $\{R|\mathbf{v}\}$, using eqn. (1.1), for each of the general equivalent positions given above. They are

$$\{E|000\} \quad \{\sigma_x|\tfrac{1}{2}00\} \quad \{\sigma_y|0\tfrac{1}{2}0\} \quad \{C_{2z}|\tfrac{1}{2}\tfrac{1}{2}0\}$$
$$\{I|000\} \quad \{C_{2x}|\tfrac{1}{2}00\} \quad \{C_{2y}|0\tfrac{1}{2}0\} \quad \{\sigma_z|\tfrac{1}{2}\tfrac{1}{2}0\}$$

where the point-group symbols used have the following meaning :

C_{2m} two-fold rotation about m axis ($m=x, y, z$),
σ_m reflection in plane normal to m axis ($m=x, y, z$),
I space inversion.

In addition to the operations just listed one can of course have any translational symmetry operation of the relevant Bravais lattice. The symbols $\{R|\mathbf{v}\}$ for all the 230 space groups have been identified by various authors (Lyubarskii 1960, Faddeyev 1964, Kovalev 1965 a, Koptsik 1966, Miller and Love 1967, Bradley and Cracknell 1972), while a useful set of rules for identifying the nature, orientation and location of each symmetry element from the equivalent general positions in the *International tables* is given in a paper by Wondratschek and Neubüser (1967).

There is a simple but important rule for the multiplication of the Seitz space-group symbols

$$\{R_2|\mathbf{v}_2\}\{R_1|\mathbf{v}_1\} = \{R_2R_1|\mathbf{v}_2 + R_2\mathbf{v}_1\} \tag{1.2}$$

as can easily be verified by applying $\{R_2|\mathbf{v}_2\}$ to both sides of eqn. (1.1). Equation (1.2) gives the rule for forming the product of two operations in connection with satisfying the various conditions involved in the abstract mathematical definition of a group.

The classical theory of symmetry is a three-dimensional study, that is the position of an atom or ion can be specified by a vector $\mathbf{r}$ $(= (x, y, z))$, in a three-dimensional Euclidean space, and one considers the effect of symmetry operations on this point. With the introduction of neutron-diffraction techniques and the consequent discovery of various new kinds of magnetic ordering in solids, it has come to be realized that to describe adequately the structure of a magnetically ordered crystal it is necessary to introduce some new developments in the theory of space groups. In recent years there have been many studies of different forms of 'generalized' symmetry, of which perhaps the best known are concerned with the 'black and white' or 'Heesch–Shubnikov' groups. Much of this work has been performed by Soviet authors and many references are to be found in recent volumes of their journal *Kristallografiya* (English translation : *Soviet Physics Crystallography*), see also Belov and Kuntsevich (1971), Bülow *et al.* (1971), Neubüser *et al.* (1971), and Wondratschek *et al.* (1971). In this article we shall confine our attention to those generalizations of classical symmetry which seem to be particularly relevant to the description of the structure of a crystal in which there exists a spontaneous ordering of some physical property associated with the atoms or ions in the crystal ; we have in mind especially magnetic ordering but also include ferroelectricity.

A convenient description of classical symmetry may be obtained by considering a scalar density function $\rho(\mathbf{r})$ which defines the probability $\rho(\mathbf{r}) \, d\mathbf{r}$ of finding an atom in the small volume element $d\mathbf{r}$ at $\mathbf{r}$. If there are several different species of atoms there will be a separate density function $\rho^p(\mathbf{r})$ for each species, p, of atom. The symmetry of the assembly of atoms in a crystal, which is described by one of the 230 classical space groups, will also be the symmetry of the density function. The effect of an arbitrary space-group operation $\{R_i|\mathbf{v}_i\}$ on the density function will be given by

$$\{R_i|\mathbf{v}_i\}\rho(\mathbf{r}) = \rho(\{R_i|\mathbf{v}_i\}^{-1}\mathbf{r}). \tag{1.3}$$

If $\{R_i|\mathbf{v}_i\}$ is a symmetry operation of the assembly of atoms in the crystal the density function will be invariant under $\{R_i|\mathbf{v}_i\}$ so that

$$\{R_i|\mathbf{v}_i\}\rho(\mathbf{r}) = \rho(\{R_i|\mathbf{v}_i\}^{-1}\mathbf{r}) = \rho(\mathbf{r}). \tag{1.4}$$

In seeking to generalize the classical ideas we have to consider the possibility of describing not just the positions of all the atoms in the crystal but also including a description of some physical property that may be possessed by these atoms.

Suppose that $A(\mathbf{r})$ is a tensor function which describes a spatial array of points in three-dimensional space, where attached to each point $\mathbf{r}$ is a tensor $A(\mathbf{r})$ which describes some physical property of an atom or ion situated at that point. A symmetry operation α_i of a magnetic crystal will involve one rotation, $R_i^{\mathbf{r}}$, for the position vector $\mathbf{r}$ of an atom, another rotation, R_i^{A}, for the tensor property, and a translation $\mathbf{v}_i$. Such an operation can be denoted by $\{R_i^{\mathrm{A}}|R_i^{\mathbf{r}}|\mathbf{v}_i\}$ and its effect on the tensor function $A(\mathbf{r})$ is given by

$$\{R_i^{\mathrm{A}}|R_i^{\mathbf{r}}|\mathbf{v}_i\}A(\mathbf{r}) = R_i^{\mathrm{A}}A(\{R_i^{\mathbf{r}}|\mathbf{v}_i\}^{-1}\mathbf{r}) = A(\mathbf{r}). \tag{1.5}$$

By analogy with eqn. (1.2) the multiplication rule for these generalized Seitz symbols $\{R_i^{\mathrm{A}}|R_i^{\mathbf{r}}|\mathbf{v}_i\}$ is

$$\{R_2^{\mathrm{A}}|R_2^{\mathbf{r}}|\mathbf{v}_2\}\{R_1^{\mathrm{A}}|R_1^{\mathbf{r}}|\mathbf{v}_1\} = \{R_2^{\mathrm{A}}R_1^{\mathrm{A}}|R_2^{\mathbf{r}}R_1^{\mathbf{r}}|\mathbf{v}_2 + R_2^{\mathbf{r}}\mathbf{v}_1\}. \tag{1.6}$$

To describe the structure of a magnetically ordered crystal we need to specify the position and the magnetic moment of each atom or ion in a unit cell of the crystal ; this requires six coordinates x, y, z, S_x, S_y and S_z, which we can write as a six-dimensional vector. It should be noted that $\mathbf{r}$ $(= (x, y, z))$ is a polar vector while $\mathbf{S}$ $(= (S_x, S_y, S_z))$ is an axial vector ; that is, $I\mathbf{r} = -\mathbf{r}$ but $I\mathbf{S} = +\mathbf{S}$, where I is the operation of space inversion. The tensor function $A(\mathbf{r})$ for a magnetically-ordered crystal can then be written as

$$A(\mathbf{r}) = \rho^p(\mathbf{r})\mathbf{S}^p(\mathbf{r}) \tag{1.7}$$

where $\rho^p(\mathbf{r})$ indicates whether there is an atom of species p at the point $\mathbf{r}$ while $\mathbf{S}^p(\mathbf{r})$ indicates the magnetic moment of such an atom. $\mathbf{S}^p(\mathbf{r})$ obviously only needs to be specified at points which are actually occupied by atoms.

When $A(\mathbf{r})$ takes the form in eqn. (1.7) the operations $\{R_i^{\mathbf{S}}|R_i^{\mathbf{r}}|\mathbf{v}_i\}$ form a group which we shall describe as a ' spin-space group '. The term ' spin-space group ' was introduced by Brinkman and Elliott (1966 a, b) with a slightly more restricted meaning and the above extension of the concept is due to Litvin (1973). Litvin describes these groups simply as ' spin groups ' and writes the symbol $\{R_i^{\mathbf{S}}|R_i^{\mathbf{r}}|\mathbf{v}_i\}$ as $[\mathscr{R}\|R\|\mathbf{v}]$. In general for a magnetically-ordered crystal there is no reason to suppose that $R_i^{\mathbf{r}}$ and $R_i^{\mathbf{S}}$ will be the same rotation. Thus $R_i^{\mathbf{r}}$ must be ' crystallographic ', by which we mean that the Seitz symbol $\{R_i^{\mathbf{r}}|\mathbf{v}_i\}$ must be an operation of one of the 230 classical, or Fedorov, space groups. The space group $\mathbf{G_0}$ formed by the operations $\{R_i^{\mathbf{r}}|\mathbf{v}_i\}$ is the classical space group that would be used to describe the symmetry of the positions of the atoms or ions in the crystal if the magnetic moments were ignored. However, $R_i^{\mathbf{S}}$ need not be restricted to being crystallographic and may be any element of the three-dimensional rotation group $O(3)$. The allowed rotations $R_i^{\mathbf{S}}$ are determined by the magnetic ordering pattern. Thus, for example, for a crystal with a simple ferromagnetic structure the magnetic moments of all the magnetic atoms or ions are aligned parallel to one fixed direction specified by a unit vector $\mathbf{n}$; the operations $R_i^{\mathbf{S}}$ then constitute the axial rotation group $\infty/m = \infty \otimes \overline{\mathrm{I}}(C_\infty \otimes C_i)$

of all possible rotations about this particular direction **n**. We may include the operation of time-inversion, θ. The position vector **r** is invariant under θ, but the effect of θ on **S** is commonly taken as being to change the sign of **S** (see, for example, the discussion in chapter 26 of the book by Wigner (1959)). If one includes θ, then the order of the group of $R_i{}^\mathbf{S}$ will be increased. For example, for a simple ferromagnetic crystal the operations $R_i{}^\mathbf{S}$ will constitute the non-unitary group $\infty/mm' = 2' \otimes \bar{1}$, instead of ∞/m. The application of spin groups to the description of more complicated magnetic ordering patterns is illustrated by Litvin (1973) for the example of the ' umbrella ' spin arrangement of CrSe.

In practice, however, for materials with certain of the important simple magnetic-ordering patterns it is not necessary to use the approach described above involving the full generality of a spin-space group, or spin group, $\mathbf{G_S}$, in six dimensions. We have stressed that in the general case of a spin-space group we do not impose any relationship between $R_i{}^\mathbf{r}$ and $R_i{}^\mathbf{S}$. If one introduces the restriction that $R_i{}^\mathbf{r}$ and $R_i{}^\mathbf{S}$ must be the *same* rotation a spin-space group reduces to a more familiar form of generalized space group, namely a ' Heesch–Shubnikov ' (or ' black and white ' or ' magnetic ') space group, **M**. An element of a Heesch–Shubnikov space group could therefore be written in the form $\{R_i | R_i | \mathbf{v}_i\}$ where we have dropped the superscripts on $R_i{}^\mathbf{S}$ and $R_i{}^\mathbf{r}$ since the two rotations, acting respectively on **S** and **r** are now the same operation. This expression for the element of a Heesch–Shubnikov space group is usually simplified to the form $\{R_i | \mathbf{v}_i\}$ which is superficially the same as for a classical space group. There is however one important difference between the use of the symbol $\{R_i | \mathbf{v}_i\}$ for classical and for Heesch–Shubnikov groups. For a classical space group the rotation R_i may be a proper rotation $C_{\mathbf{n}\alpha}$, through an angle α about an axis pointing in the direction of the unit vector **n**, or it may be an improper rotation $IC_{\mathbf{n}\alpha}$. For a Heesch–Shubnikov space group R_i may be $C_{\mathbf{n}\alpha}$ or $IC_{\mathbf{n}\alpha}$, or it may involve θ and be of the form $\theta C_{\mathbf{n}\alpha}$ or $\theta IC_{\mathbf{n}\alpha}$. We do not intend to give a lengthy exposition of the subject of Heesch–Shubnikov groups because several other reviews exist already which cover the subject quite comprehensively, see, for example, Birss (1964), Opechowski and Guccione (1965), Bradley and Davies (1968), Cracknell (1969 e, 1974 a) and chapter 7 of Bradley and Cracknell (1972).

As an example of the use of the Heesch–Shubnikov space groups in the description of simple magnetic structures we illustrate the unit cell of antiferromagnetic MnF_2 in fig. 1. Above its Néel temperature the symmetry of this material, which has the rutile structure, is described by the classical space group $P4_2/mnm$. Below the Néel temperature we have $\mathbf{S} = (0, 0, S)$ for the Mn atoms of one sublattice and $\mathbf{S} = (0, 0, -S)$ for the Mn atoms on the other sublattice and the symmetry of the structure is described by the Heesch–Shubnikov space group $P4_2'/mnm'$ (for full details see Dimmock and Wheeler (1962 a)). In the more abstract formal development of the Heesch–Shubnikov groups using a four-dimensional space the first three dimensions are taken as the three classical Euclidean space dimensions of x, y and z and the fourth dimension involves a coordinate ϕ that is only allowed to take one of two values, such as either $+$ and $-$ or black and white in the common treatments. The use of the concept of colour, and of the idea of anti-symmetry, is an extremely valuable pedagogic device for use in the

Fig. 1

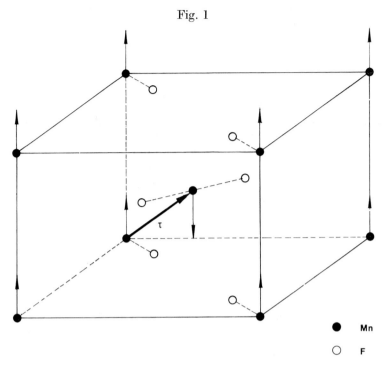

Mn

F

Unit cell of antiferromagnetic MnF_2.

Fig. 2

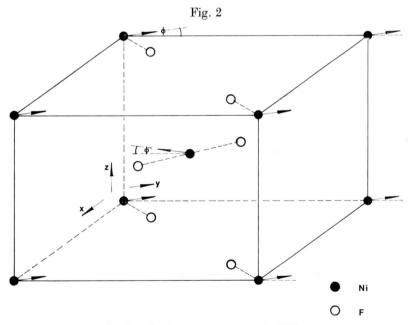

Ni

F

Unit cell of antiferromagnetic NiF_2.

exposition of the formal derivation and enumeration of all the Heesch–Shubnikov groups. In the case of MnF_2 the Heesch–Shubnikov group which describes the symmetry of the structure can conveniently be regarded as acting in a four-dimensional vector space, instead of a six-dimensional vector space, because two components of **S** vanish and so there is a close similarity with the more abstract formal development of the Heesch–Shubnikov groups as 'black and white' groups. However, it is important to emphasize that the use of the Heesch–Shubnikov groups in the description of magnetic structures is not necessarily restricted to a four-dimensional vector space. Once the Heesch–Shubnikov groups are derived and one comes to use them in describing magnetic structures, it is perhaps a good idea to forget about the idea of 'colour' in this connection. We can illustrate this by the example of antiferromagnetic NiF_2 which also has the rutile structure above its Néel temperature. Below its Néel temperature this material has a canted antiferromagnetic structure which is illustrated in fig. 2. The atoms at the corners of the unit cell illustrated in fig. 2 are on one sublattice and for each of these atoms $\mathbf{S} = (-S \sin \delta, S \cos \delta, 0)$, while the atom at the centre of this unit cell is on the other sublattice and has $\mathbf{S} = (-S \sin \delta, -S \cos \delta, 0)$. The symmetry of this structure can be described by the Heesch–Shubnikov space group $Pnn'm'$ which contains the following symmetry operations :

$$\{E|000\}, \quad \{C_{2x}|\tfrac{1}{2}\tfrac{1}{2}\tfrac{1}{2}\}, \quad \{I|000\}, \quad \{\sigma_x|\tfrac{1}{2}\tfrac{1}{2}\tfrac{1}{2}\},$$

$$\theta\{C_{2z}|000\}, \quad \theta\{C_{2y}|\tfrac{1}{2}\tfrac{1}{2}\tfrac{1}{2}\}, \quad \theta\{\sigma_z|000\}, \quad \theta\{\sigma_y|\tfrac{1}{2}\tfrac{1}{2}\tfrac{1}{2}\}.$$

For NiF_2 the Heesch–Shubnikov space group can be regarded as acting in a five-dimensional vector space, instead of a six-dimensional vector space because the z component of **S** vanishes on both sublattices, but in this case the relationship with the development as 'black and white' groups in a four-dimensional vector space is more obscure.

As an alternative to constructing new groups to describe a magnetically ordered structure, it is possible to describe such a structure in terms of the irreducible representations of the classical space group, $\mathbf{G}_0$, which describes the symmetry of the atomic positions. This procedure, which is known as 'representation analysis' and which is particularly useful in the interpretation of experimental data for structure determination, will be discussed in § 8.1.

1.3. *Basic principles*

The application of group theory to solid-state physics is based on two results. The first of these is *Neumann's principle* which is essentially a classical or macroscopic feature of the physics of a crystalline solid. The second is *Wigner's theorem* and this is a result which arises in connection with the quantum-mechanical description of a system that possesses a certain symmetry. These two results are associated with two rather different approaches to the idea of the symmetry of a system. In the first approach one considers that group of symmetry operations under which the system is invariant. This is the approach which is adopted in choosing a point group or space group for the description of the symmetry of a given crystal ; it is this approach which is implied in the application of Neumann's principle.

The second approach is that of the quantum-mechanical physicist who constructs a Hamiltonian which is invariant under the operations of some point group or space group **G**. Quite often the group of the Hamiltonian is taken to be the point group or space group which describes the macroscopic symmetry of the system, but this need not necessarily be so. However, as far as the wave function $\psi(\mathbf{r})$ is concerned, what is of interest from the symmetry point of view is the transformation properties of $\psi(\mathbf{r})$ under all the operations of **G**; this is the approach which is implied in the application of Wigner's theorem. It is not usually very profitable to consider that subgroup of **G** which consists of only those operations that leave $\psi(\mathbf{r})$ invariant.

Suppose that the symmetry operations R of a certain crystal form one of the (crystallographic) point groups **G**. By saying that R is a symmetry operation, or a covering operation, of the crystal we mean that the appearance of the crystal after the operation R has been performed is indistinguishable from its appearance before the operation was performed. In a simple-minded study of the point-group symmetry of a crystal the term ' appearance ' would refer to the orientations of the faces, or of the face normals, of the crystal. Neumann's principle is an extension of this to the idea that one cannot distinguish between the two states of the crystal, before and after the performance of the symmetry operation R, by studying *any* of the physical properties of these two states. Stated formally it would be (see, for example, Nye (1957, p. 20) or Birss (1964, p. 44)):

Neumann's principle

' The symmetry operations of any physical property of a crystal must include the symmetry operations of the point group of the crystal.'
Some of the more direct consequences of Neumann's principle will be noted in §§ 3.1 and 7.1. The principle was originally enunciated only for non-magnetic crystals but it can be extended to the case of magnetically ordered crystals as well and this will be discussed in §§ 3.2 and 3.3.

In the present article we are assuming that the reader has a general familiarity, in outline at least, with the concept of a group in the mathematical sense. We are also assuming that the reader is aware of the existence and importance of the idea of the representation of an abstract group by groups of unitary matrices. We recall that a representation Γ of a group **G** consists of a set of matrices $\mathbf{D}(R_i)$ where some rule is given for establishing the correspondence between $\mathbf{D}(R_i)$ and R_i, the elements of the abstract group. The set of matrices must themselves form a group and must also satisfy the following condition :

$$\mathbf{D}(R_j)\mathbf{D}(R_k) = \mathbf{D}(R_j R_k) \qquad (1.8)$$

for all possible pairs of elements R_j and R_k in **G**. For practical purposes the representations of a group which are usually of interest are those representations which are *irreducible*. Neumann's principle can be expressed in terms of representation theory by saying that any physical property of a crystal must belong to the identity representation Γ_1 (or Γ_1^+) of the point group, **G**, of the crystal. We also recall that a basis of a representation may be taken as a row vector $\langle \mathbf{u}|$, with components u_p, which transforms in the following manner under an operation R_j of the group **G** :

$$R_j u_p = \sum_{q=1}^{d} u_q \mathbf{D}(R_j)_{qp} \tag{1.9}$$

where d is the dimension of the representation. The construction of functions which form a basis of a given irreducible representation of a point group or space group is an important step in several of the arguments involved in the application of group theory. This can be done by starting with some arbitrary function u and generating the functions

$$u_p = W_{pr} u = \frac{d}{|\mathbf{G}|} \sum_{R \in \mathbf{G}} \mathbf{D}(R)_{pr}{}^* R u. \tag{1.10}$$

Provided that u_p does not vanish the functions $\langle u_p |$ $(p = 1, 2, ..., d)$ form a basis of Γ and u_p is sometimes described as a function which is *symmetry-adapted* to the row p of Γ. The operator $W_{pp}{}^i$ is sometimes called a *projection operator* ; it is sometimes replaced by

$$(d/|\mathbf{G}|) \sum_{R \in \mathbf{G}} \chi(R)^* R.$$

Suppose now that one is concerned with a microscopic or quantum-mechanical approach to the study of a system which possesses the symmetry of a point group **G**. Then, naturally, one will be concerned with the form of the wave function of the system. But it is generally accepted that a wave function, $\psi(\mathbf{r})$, is not a ' physically observable property ' of a system, although of course it can be used in the determination of matrix elements which can then be related to various physical properties of the system. Thus one cannot apply Neumann's principle to $\psi(\mathbf{r})$ and assume that $\psi(\mathbf{r})$ will possess all the symmetry of the point group **G**. However, the point-group symmetry of the system will impose some restrictions on the allowed forms of the wave function $\psi(\mathbf{r})$ of that system. The Hamiltonian operator, $\mathscr{H}$, of the system can be regarded as having the symmetry of some crystallographic point group or space group. The restrictions on the form of $\psi(\mathbf{r})$ arise out of Wigner's theorem which can be stated as follows:

Wigner's theorem

' If R is a symmetry operation of the Hamiltonian operator, $\mathscr{H}$, which describes a quantum-mechanical system and if $\psi(\mathbf{r})$ is an eigenfunction of $\mathscr{H}$, then $R\psi(\mathbf{r})$ is also an eigenfunction of $\mathscr{H}$ and has the same eigenvalue E as $\psi(\mathbf{r})$.'

The theorem is not restricted to point groups but also applies to space groups as well. What the theorem means in practical terms is that, for a system belonging to the point group or space group **G**, any wave function $\psi(\mathbf{r})$ of the system must belong to, that is must transform according to, one or other of the irreducible representations Γ_i of **G**. That is, the wave function $\psi(\mathbf{r})$ of a particle or quasiparticle in a molecule or crystal must be one component of a basis of one of the irreducible representations of the point group or space group **G** of that molecule or crystal. Therefore, the transformation properties of $\psi(\mathbf{r})$ under the symmetry operations R can be determined by using the matrices of the appropriate irreducible representation of

G. This can be used to simplify the form of $\psi(\mathbf{r})$ that is used in various calculations. Moreover, the degeneracies of the energy levels can be predicted because they are determined by the degeneracies of the irreducible representations of **G**. Many of the sections in this article will be concerned with examples of the use of Wigner's theorem.

If the symmetry of a system is described by one of the Heesch–Shubnikov groups, instead of by one of the classical groups, it is necessary to make some modification to the discussion that we have just given. Each Heesch–Shubnikov group which contains elements involving θ, the operation of anti-symmetry, possesses a halving subgroup of elements that do not involve θ. This halving subgroup will be one of the classical point groups or space groups. This can be seen from the symmetry operations identified above in § 1.2 for NiF_2. Thus a Heesch–Shubnikov point group (or space group) can be written in the form

$$\mathbf{M} = \mathbf{H} + A\,\mathbf{H} \tag{1.11}$$

where **H** is one of the classical point groups (or space groups) and A is some operation of the form θR (or $\{\theta S\,|\,\mathbf{w}\}$). If it is possible to choose A to be θ, the operation of anti-symmetry, by itself, then **M** can be regarded as the direct product $\mathbf{H} \otimes (E + \theta)$; in this case **M** is sometimes described as a grey point group (or space group) (see, for example, Cracknell (1966 b), Bradley and Davies (1968), or § 7.2 of Bradley and Cracknell (1972)).

It is possible to show that the operation of anti-symmetry θ, when considered as the operation of time-inversion, is anti-unitary instead of unitary, that is

$$\theta\Big[\sum_{k} a_{k}\psi_{k}\Big] = \sum_{k} a_{k}{}^{*}\theta\psi_{k} \tag{1.12}$$

instead of giving $\sum_{k} a_{k}\theta\psi_{k}$. The details of the proof of this property do not concern us here (see, for example, chapter 26 of the book by Wigner (1959)) but it has some important consequences. In particular, it is possible to show that for a group containing anti-unitary elements one cannot find a group of matrices which satisfy eqn. (1.11). In other words it is not possible to construct matrix representations of **M**. However, it is possible to modify the theory of representations and to obtain sets of matrices which form what are called *co-representations* of **M**. The conditions satisfied by a set of matrices which form a co-representation of the non-unitary group **M** $(=(\mathbf{H} + A\,\mathbf{H}))$ are given by

$$\left.\begin{array}{l} \mathbf{D}(R_{j})\mathbf{D}(R_{k}) = \mathbf{D}(R_{j}R_{k}) \\[2mm] \mathbf{D}(R_{j})\mathbf{D}(A_{k}) = \mathbf{D}(R_{j}A_{k}) \\[2mm] \mathbf{D}(A_{j})\mathbf{D}(R_{k})^{*} = \mathbf{D}(A_{j}R_{k}) \\[2mm] \mathbf{D}(A_{j})\mathbf{D}(A_{k})^{*} = \mathbf{D}(A_{j}A_{k}) \end{array}\right\} \tag{1.13}$$

where R_{j} and R_{k} are elements from the unitary subgroup **H** and A_{j} and A_{k} are elements from the set $A\,\mathbf{H}$ of anti-unitary operations of **M**. Co-representations of non-unitary groups may be either reducible or irreducible in the same way that representations of unitary groups may be either reducible or irreducible. For practical purposes it is nearly always the irreducible co-representations which are of importance. It is possible to determine all the irreducible co-representations of a non-unitary group **M**, given by eqn. (1.11), if all the irreducible representations of **H**, the halving subgroup of unitary elements, are already known (for details see Wigner (1959), Dimmock and Wheeler (1962 b, 1964), Dimmock (1963), Bradley and Davies (1968), Cracknell (1969 e), Bradley and Cracknell (1972)). We shall not reproduce the details here but simply note that an irreducible representation, Γ_i, of the unitary subgroup **H** may lead to an irreducible co-representation of **M** in one of three different ways, which are referred to as case (a), case (b), and case (c). In case (a) the representation Γ_i leads to a co-representation $D\Gamma_i$ of **M** of dimension equal to the dimension of Γ_i. In case (b) Γ_i leads to a co-representation $D\Gamma_i$ of **M** of dimension twice the dimension of Γ_i. In case (c) two non-equivalent irreducible representations Γ_i and Γ_j, which have the same dimension, 'stick together' to form an irreducible co-representation $D\Gamma_{i,j}$ of **M** with dimension twice that of either Γ_i or Γ_j. For a magnetic crystal Wigner's theorem means that the symmetry properties of the eigenfunctions of the Hamiltonian $\mathscr{H}$ will be determined by the irreducible co-representations of **M** and the degeneracies of the eigenvalues will be determined by the degeneracies of these irreducible co-representations. In fig. 3 we illustrate the relationship between the pattern of the energy levels for a system which has only the symmetry of **H** and for the same system when the anti-unitary symmetry operations of A**H** are included.

Fig. 3

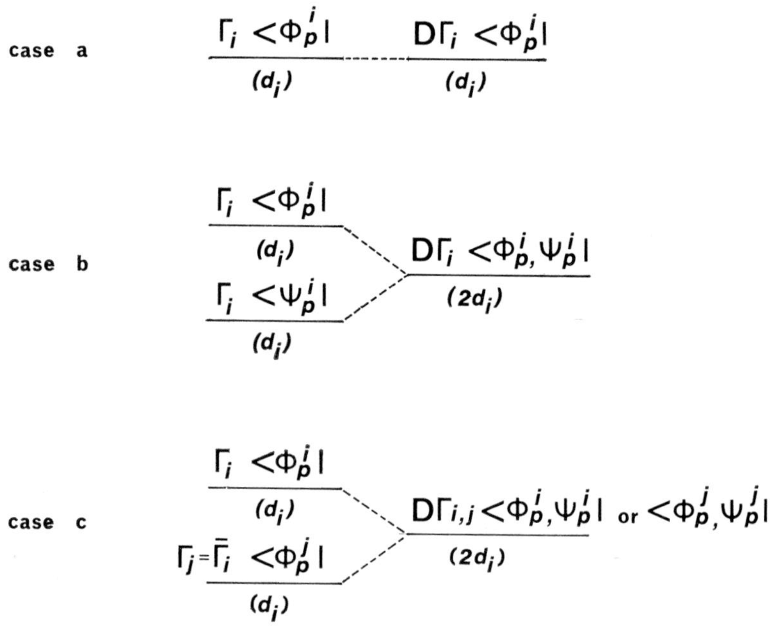

Possible changes in the degeneracies of eigenfunctions and eigenvalues when the anti-unitary elements A**H** are added to the subgroup **H** of **M**. $(\langle \psi_p{}^i | = A \langle \phi_p{}^i |.)$

1.4. *Representation theory for point groups and space groups*

The discussion of Wigner's theorem in the previous section will have made it clear that the irreducible representations of a group play a very important rôle in the applications of group theory in solid-state physics. It would not be appropriate in the present work to engage in a lengthy discussion of the details of the theory of the representations of point groups and space groups ; it is assumed that the general outlines of the theory are either already familiar to the reader or can be learned from the existing literature (see, for example, Heine (1960), Tinkham (1964), Cornwell (1969), Bradley and Cracknell (1972)). We shall simply note a few of the salient points in the theory by way of revision and also to establish the notation that we shall use in later sections. It would also not be appropriate to include extensive tables of representations of point groups and space groups in this article ; we shall simply give references to indicate where these tables may be found. We shall just include, where necessary, extracts from the tables for those point groups or space groups which describe the symmetries of the particular structures that we shall use in our examples. To avoid excessive tabular material we shall try, where possible, to use the same structure to illustrate several different applications.

Each of the classical point groups is a finite group of fairly low order. The irreducible representations of these groups were determined a long time ago and they have been published in many different articles and books ; it would be wasteful to reproduce them again here. Instead of giving the complete matrices $\mathbf{D}(R_i)$ for the elements R_i in a point group $\mathbf{G}$, where these matrices satisfy the condition in eqn. (1.8) as well as the condition of irreducibility, it is common only to give the characters of these matrices. For many practical purposes it is quite adequate, as well as being much more convenient, to use just the characters and not the complete matrices ; some examples will be found in tables 2 and 17. Many different sets of notation have been used, both in labelling the symmetry operations in the point groups and also in labelling the irreducible representations themselves. We shall follow the notation of Bradley and Cracknell (1972) in labelling the symmetry operations for the point groups, while for labelling the irreducible representations of the point groups we shall use the notation given by Koster *et al.* (1963) which was also reproduced in the book by Bradley and Cracknell (1972). The use of singular matrix representations of point groups has been suggested by Killingbeck (1972) (but see also Bell (1972) and Subramanian (1973)).

At the end of § 1.3 we have noted that there is a simple relationship between the irreducible co-representations of a black and white, or magnetic, point group

$$\mathbf{M} = \mathbf{H} + A\,\mathbf{H} \qquad (1.11)$$

and the irreducible representations of the unitary subgroup, $\mathbf{H}$. Thus it is not necessary to determine and tabulate the irreducible co-representations of $\mathbf{M}$ completely. It is adequate, for each $\mathbf{M}$, to identify $\mathbf{H}$, A, and the irreducible representations Γ_i of $\mathbf{H}$; $\mathbf{H}$, of course, is one of the 32 classical point groups. For each $\mathbf{M}$ it then only remains to identify the case ((a), (b), or (c)) to which each Γ_i belongs and to identify the matrix of the additional generating element A for the co-representation derived from Γ_i. Tables identifying the irreducible co-representations of the 58 black and white, or magnetic, point groups have been published (Dimmock and Wheeler 1962 b, Cracknell 1966 b, Bradley and Cracknell 1972); an example is included later in table 17.

The determination of the irreducible representations of a space group is more difficult than determining the irreducible representations of a point group. This is because a space group is a group of infinite order. The solution of the problem of determining the irreducible representations of a space group has its origins in the work of Seitz (1936). The first application was to the labelling and the study of the degeneracies of the energy bands in face-centred cubic, body-centred cubic, and simple cubic crystals by Bouckaert et al. (1936) (see § 2.2). The key to the determination of the irreducible representations of a space group is to be found in the wave vector $\mathbf{k}$. Moreover, once these representations have been determined, the scheme that is used for labelling them is also based on the use of the wave vector $\mathbf{k}$.

To determine the irreducible representations of a three-dimensional space group it is conventional to consider a crystal of finite extent and then to apply periodic boundary conditions. If $\mathbf{t}_1$, $\mathbf{t}_2$ and $\mathbf{t}_3$ are the basic vectors of the Bravais lattice of the crystal then the space group $\mathbf{G}$ of the crystal has a subgroup of translations $\{E|\mathbf{T_m}\}$ where $\mathbf{T_m}$ is of the form

$$\mathbf{T_m} = m_1\mathbf{t}_1 + m_2\mathbf{t}_2 + m_3\mathbf{t}_3 \qquad (1.14)$$

and where m_1, m_2 and m_3 are integers. The reciprocal lattice vectors $\mathbf{g_1}$, $\mathbf{g_2}$ and $\mathbf{g_3}$ may be defined by

$$\mathbf{g}_i \cdot \mathbf{t}_j = 2\pi\delta_{ij} \qquad (1.15)$$

and can be written down explicitly in terms of $\mathbf{t}_1$, $\mathbf{t}_2$ and $\mathbf{t}_3$ if required. We suppose that there are N_i unit cells ($i = 1, 2, 3$) in the direction $\mathbf{t}_i$ in the crystal where N_i is a very large number. By applying the Born–von Kármán cyclic boundary conditions which can be expressed in terms of Seitz space-group symbols as

$$\{E|l_1N_1\mathbf{t}_1 + l_2N_2\mathbf{t}_2 + l_3N_3\mathbf{t}_3\} = \{E|\mathbf{0}\} \qquad (1.16)$$

where l_1, l_2 and l_3 are integers, one obtains a finite group. The application of Bloch's theorem then shows that the wave vector

$$\mathbf{k} = k_1\mathbf{g}_1 + k_2\mathbf{g}_2 + k_3\mathbf{g}_3 \qquad (1.17)$$

forms a good quantum number for labelling particle or quasiparticle states in the crystal. That is, three-dimensional periodicity in the crystal, or more precisely in the Hamiltonian of the problem, leads to the classification of quantum-mechanical eigenstates by a wave vector $\mathbf{k}$ which is also in a three-dimensional space. All the physically distinguishable eigenstates of the Hamiltonian operator of a crystal can be assigned to wave vectors within a Brillouin zone, which is just a unit cell of $\mathbf{k}$ space. The particular type of unit cell of the reciprocal lattice that is usually chosen as the Brillouin zone is the Wigner–Seitz unit cell. For crystals of high symmetry this presents no particular difficulty and, in such crystals, the labels assigned to the special

points of symmetry and lines of symmetry are quite generally accepted.†
However, for crystals of low symmetry there is no uniformity either in the
choice of the shape of the Brillouin zone or in the labels assigned to the special
points of symmetry and lines of symmetry. Brillouin zones for all the space
groups are illustrated by a number of authors (Koster 1957, Meijer and Bauer
1962, Bradley and Cracknell 1972). Examples of Brillouin zones for various
structures will be encountered later, see figs. 5, 6, 8, 13, 14 and 21.

It is not generally appreciated that there is a lack of uniqueness in the
choice of the Brillouin zone of any given crystal. Indeed, it has been strongly
argued elsewhere (see, for example, Bradley and Cracknell (1972, pp. 90–2)
that for crystals of very low symmetry, namely monoclinic and triclinic
crystals, it is more convenient to use the primitive unit cell, instead of the
Wigner–Seitz unit cell, of the reciprocal lattice to define the Brillouin zone.
A primitive unit cell is a parallelepiped centred at one reciprocal lattice point
and having edges $\mathbf{g}_1$, $\mathbf{g}_2$ and $\mathbf{g}_3$, where $\mathbf{g}_1$, $\mathbf{g}_2$ and $\mathbf{g}_3$ are the basis vectors of the
reciprocal lattice. One important consequence of the lack of uniqueness in
the choice of that unit cell of the reciprocal lattice which one intends to use
as the Brillouin zone is the fact that the distinction between normal processes
and umklapp processes is not unique either, but depends on one's choice of
Brillouin zone. This is illustrated schematically for a two-dimensional
crystal in fig. 4. Suppose an electron with wave vector $\mathbf{k}_1$ absorbs a phonon
with wave vector $\mathbf{k}_2$ and thereby acquires a final wave vector $\mathbf{k}_3$ or $\mathbf{k}_3'$. With
the choice of Brillouin zone in fig. 4 (*a*) this would be described as an umklapp
process but with the alternative choice of Brillouin zone in fig. 4 (*b*) it would
be described as a normal process. Although in the past those writers who are
careful about this matter have appreciated that a change in the choice of the
Brillouin zone will alter the division into normal processes and umklapp
processes, there are many examples of careless writing, indicating that some
authors are not fully aware of this point. As indicated already, this difficulty
only arises in practice for crystals of rather low symmetry.

For any space group $\mathbf{G}$ the irreducible representations of $\mathbf{G}$ associated
with a given wave vector $\mathbf{k}$ can be regarded, for most practical purposes, as
the irreducible representations $\Gamma_j{}^\mathbf{k}$ of the *little group* $\mathbf{G}^\mathbf{k}$. $\mathbf{G}^\mathbf{k}$ is a subgroup
of $\mathbf{G}$ and contains all the operations $\{R|\mathbf{v}\}$ of $\mathbf{G}$ which satisfy the condition
that $R\mathbf{k}$ is equivalent to $\mathbf{k}$, that is, $R\mathbf{k}$ is equal to $\mathbf{k}$ or only differs from $\mathbf{k}$
by a translation vector $(n_1\mathbf{g}_1 + n_2\mathbf{g}_2 + n_3\mathbf{g}_3)$ of the reciprocal lattice (where
n_1, n_2 and n_3 are integers). The point group that is obtained by taking all
the point-group operations R, from operations $\{R|\mathbf{v}\}$ of $\mathbf{G}$, for which $R\mathbf{k}$ is
equivalent to $\mathbf{k}$ is sometimes denoted by $\mathbf{P}(\mathbf{k})$. For a symmorphic space
group, such as $Fm3m$ $(O_h{}^5)$ for example (see § 2.2), the irreducible representations of $\mathbf{G}^\mathbf{k}$ are rather trivially related to the irreducible representations of
the point group $\mathbf{P}(\mathbf{k})$. We should, perhaps, repeat the definitions of the
terms points, lines, and planes of symmetry to avoid any possible misunder-
standings (see Bradley and Cracknell (1972)).

---

† There are actually two systems in use for labelling the special points of symmetry
and lines of symmetry in the Brillouin zone. In the Soviet Union they are labelled
as $\mathbf{k}_1$, $\mathbf{k}_2$, ..., $\mathbf{k}_i$, ..., whereas in the West they are labelled by a slightly bizarre
collection of capital letters from the Greek and Roman alphabets ; each of these
systems is widely accepted within its own political sphere of influence.

Fig. 4

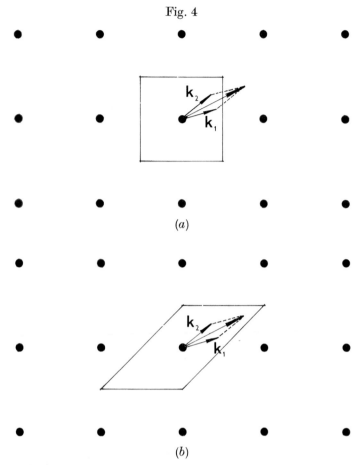

Diagram to illustrate that the distinction between normal and umklapp processes may depend on the choice of Brillouin zone. The sum $(\mathbf{k}_1 + \mathbf{k}_2)$ is outside the Brillouin zone in (a) but inside it in (b).

Definition. '$\mathbf{k}$ is a *point of symmetry* if there exists a neighbourhood N of $\mathbf{k}$ in which no point except $\mathbf{k}$ has the symmetry group $\mathbf{P}(\mathbf{k})$.'

It should be noted that K and U in fig. 8, which are sometimes regarded as points of symmetry, do not come within this definition because they possess no more symmetry than the neighbouring lines Σ and S, respectively.

Definition. 'If in any sufficiently small neighbourhood N of $\mathbf{k}$ there is always a line (plane) in N passing through $\mathbf{k}$, all points of which have the same symmetry group as $\mathbf{k}$, then $\mathbf{k}$ is said to be a *line (plane) of symmetry.*'

Finally, if in any sufficiently small neighbourhood N of $\mathbf{k}$ all points of N have the same symmetry group then $\mathbf{k}$ is said to be a *general point*.

The set of distinct (that is, non-equivalent) wave vectors $R\mathbf{k}$ which is obtained by taking all the point-group operations R, from operations $\{R|\mathbf{v}\}$ of $\mathbf{G}$, for which $R\mathbf{k}$ is not equivalent to $\mathbf{k}$, is called the '*star*' (or occasionally the '*orbit*') of $\mathbf{k}$. In determining the irreducible representations of a space group **G** it is only necessary to determine the irreducible representations $\Gamma_j^{\mathbf{k}}$ of $\mathbf{G}^{\mathbf{k}}$ for one wave vector in each star. In this connection it is convenient to recall certain definitions. These are :

(i) *Basic domain*

' For each Brillouin zone there is a *basic domain*, Ω, such that $(\sum_R R\Omega)$ is equal to the whole Brillouin zone, where R are the elements of the holo-symmetric point group of the relevant crystal system.'

(ii) *Representation domain*

' For any space group **G** there is a *representation domain*, Φ, of the appropriate Brillouin zone such that $(\sum_R R\Phi)$ is equal to the whole Brillouin zone, where the sum over R runs through the elements of the isogonal point group of **G**.'

and

(iii) *Co-representation domain*

' For a magnetic space group **M** there is a *co-representation domain*, Θ, of the appropriate Brillouin zone such that $(\sum_R R\Theta)$ is equal to the whole Brillouin zone, where the sum over R runs over all the elements of the isogonal point group **P**' of **M**.'

For a classical space group, **G**, the term ' isogonal point group ' refers to the point group obtained by taking all the operations R found among the elements $\{R|\mathbf{v}\}$ of **G**. For a non-unitary group **M** the isogonal point group is defined to be the point group consisting of all the operations S and S', where S ranges over all the elements $\{S|\mathbf{w}\}$ of **H** and S' ranges over all the elements $\theta\{S'|\mathbf{w}'\}$ of the set IA **H**. It should be noticed that although **M** is a non-unitary, or black and white, group the isogonal point group **P**' is a classical point group and not a black and white point group. In fig. 5 we illustrate a basic domain and a representation domain for one particular cubic space group, $Pm3$ (T_h^1), and in fig. 6 we illustrate co-representation domains for the space groups of antiferromagnetic NiF_2 and MnF_2. The importance of these definitions lies in the fact that for any given classical space group it is only necessary to determine the irreducible representations *ab initio* over a certain fraction of the Brillouin zone, namely the representation domain, and not over the whole of the Brillouin zone. In practice it is found that in most cases it is only necessary to determine the irreducible representations within a single basic domain which may be even smaller than the representation domain (for details see § 5.5 of Bradley and Cracknell (1972)). In a similar manner, if the symmetry of a magnetically ordered crystal is described by one of the Heesch–Shubnikov space groups it is not necessary to determine the irreducible co-representations *ab initio* all over the Brillouin zone but only over a fraction of the Brillouin zone, namely the *co-representation domain* (Cracknell 1973 b).

1.5. *Kronecker products*

An important concept that is used in several applications is that of the ' Kronecker product ' of two representations, Γ_i and Γ_j, of a point group **G**.

A. P. Cracknell

Fig. 5

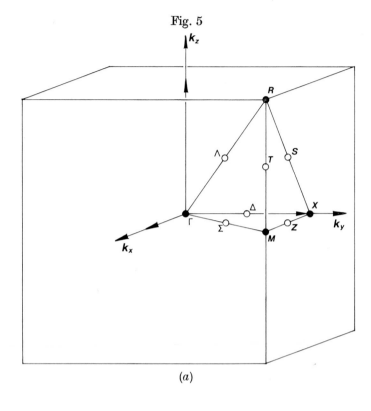

(a)

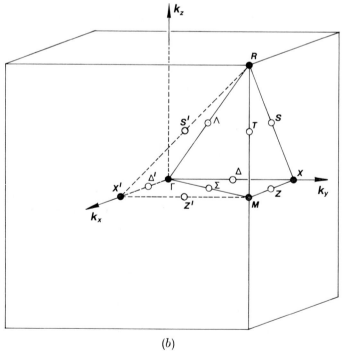

(b)

(a) Basic domain, Ω, and (b) representation domain, Φ, for the space group $Pm3$
(T_h^1).

If the row vectors $|\mathbf{u}^i\rangle$ and $|\mathbf{v}^j\rangle$, with components $u_p{}^i$ and $v_q{}^j$ respectively, form bases of the representations Γ_i and Γ_j of the point group **G**, then the row vector $|\mathbf{u}^i\mathbf{v}^j\rangle$ forms a basis of a representation which is called the Kronecker product $\Gamma_i\boxtimes\Gamma_j$ of **G**. $\Gamma_i\boxtimes\Gamma_j$ may be reducible. If we write

$$\Gamma_i\boxtimes\Gamma_j = \sum_k c_{ij,\,k}\Gamma_k \tag{1.18}$$

Fig. 6

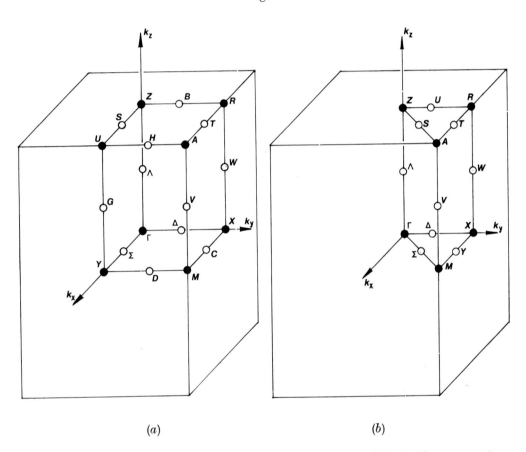

(a)   (b)

Co-representation domain for (a) antiferromagnetic NiF_2 and (b) antiferromagnetic MnF_2.

the coefficients $c_{ij,\,k}$ are called the Clebsch–Gordan coefficients. Tables giving the reductions of all the Kronecker products for each of the 32 point groups will be found, for example, in the book by Koster *et al.* (1963). As an example we reproduce in table 2 the part of those tables referring to the point group 432 ; this will be used later in this article in connection with some particular problems (see § 4.2).

A. P. Cracknell

The individual components $u_p{}^i v_q{}^j$ of the vector $|\mathbf{u}^i \mathbf{v}^j\rangle$ may not directly be bases of the irreducible representations Γ_k, because it may have been necessary to perform a unitary transformation to reduce $\Gamma_i \boxtimes \Gamma_j$. If $\mathbf{S}$ is the unitary matrix used in the reduction of $\Gamma_i \boxtimes \Gamma_j$ then the components of the transformed basis $|\mathbf{u}^i \mathbf{v}^j\rangle \mathbf{S}$ will belong to the various representations Γ_k. The components of $|\mathbf{u}^i \mathbf{v}^j\rangle \mathbf{S}$ will be linear combinations of the $u_p{}^i v_q{}^j$ terms and the particular linear combinations that are appropriate to each of the representations Γ_k are identified in the tables of coupling coefficients given by Koster *et al.* (1963). Thus, for example, suppose that $|u_1{}^3, u_2{}^3\rangle$ and $|v_1{}^5, v_2{}^5, v_3{}^5\rangle$ are bases of Γ_3 and Γ_5, respectively, of the point group 432. The reduction of the Kronecker product $\Gamma_3 \boxtimes \Gamma_5$ is given in table 2 :

$$\Gamma_3 \boxtimes \Gamma_5 = \Gamma_4 \oplus \Gamma_5. \tag{1.19}$$

Table 3 is an extract from table 83 of Koster *et al.* (1963), where $\psi_i{}^4$ is the ith component of a basis of Γ_4 and $\psi_i{}^5$ is the ith component of a basis of Γ_5. From this table we see that the bases for Γ_4 and Γ_5 constructed from the product of bases of Γ_3 and Γ_5 are

$$\Gamma_4 : \; \left| (-\tfrac{1}{2}\sqrt{3} u_1{}^3 v_1{}^5 - \tfrac{1}{2} u_2{}^3 v_1{}^5), (\tfrac{1}{2}\sqrt{3} u_1{}^3 v_2{}^5 - \tfrac{1}{2} u_2{}^3 v_2{}^5), u_2{}^3 v_3{}^5 \right\rangle \tag{1.20}$$

and

$$\Gamma_5 : \; \left| (-\tfrac{1}{2} u_1{}^3 v_1{}^5 + \tfrac{1}{2}\sqrt{3} u_2{}^3 v_1{}^5), (-\tfrac{1}{2} u_1{}^3 v_2{}^5 - \tfrac{1}{2}\sqrt{3} u_2{}^3 v_2{}^5), u_1{}^3 v_3{}^5 \right\rangle. \tag{1.21}$$

Table 2. Irreducible representations of the point group 432 (*O*)
(*a*) Character table

	C_1	C_2	C_3	C_4	C_5	C_6	C_7	C_8
Γ_1	1	1	1	1	1	1	1	1
Γ_2	1	1	1	1	1	-1	-1	-1
Γ_3	2	2	2	-1	-1	0	0	0
Γ_4	3	3	-1	0	0	1	1	-1
Γ_5	3	3	-1	0	0	-1	-1	1
Γ_6	2	-2	0	1	-1	$\sqrt{2}$	$-\sqrt{2}$	0
Γ_7	2	-2	0	1	-1	$-\sqrt{2}$	$\sqrt{2}$	0
Γ_8	4	-4	0	-1	1	0	0	0

C_1 E

C_2 $\bar{E}$

C_3 $C_{2x}, C_{2y}, C_{2z}, \bar{C}_{2x}, \bar{C}_{2y}, \bar{C}_{2z}$

C_4 $C_{31}{}^-, C_{32}{}^-, C_{33}{}^-, C_{34}{}^-, C_{31}{}^+, C_{32}{}^+, C_{33}{}^+, C_{34}{}^+$

C_5 $\bar{C}_{31}{}^-, \bar{C}_{32}{}^-, \bar{C}_{33}{}^-, \bar{C}_{34}{}^-, \bar{C}_{31}{}^+, \bar{C}_{32}{}^+, \bar{C}_{33}{}^+, \bar{C}_{34}{}^+$

C_6 $C_{4x}{}^+, C_{4y}{}^+, C_{4z}{}^+, C_{4x}{}^-, C_{4y}{}^-, C_{4z}{}^-$

C_7 $\bar{C}_{4x}{}^+, \bar{C}_{4y}{}^+, \bar{C}_{4z}{}^+, \bar{C}_{4x}{}^-, \bar{C}_{4y}{}^-, \bar{C}_{4z}{}^-$

C_8 $C_{2a}, C_{2b}, C_{2c}, C_{2d}, C_{2e}, C_{2f}, \bar{C}_{2a}, \bar{C}_{2b}, \bar{C}_{2c}, \bar{C}_{2d}, \bar{C}_{2e}, \bar{C}_{2f}$

Table 2—continued

(b) Reductions of Kronecker products

	Γ_1	Γ_2	Γ_3	Γ_4	Γ_5	Γ_6	Γ_7	Γ_8
Γ_1	Γ_1	Γ_2	Γ_3	Γ_4	Γ_5	Γ_6	Γ_7	Γ_8
Γ_2		Γ_1	Γ_3	Γ_5	Γ_4	Γ_7	Γ_6	Γ_8
Γ_3			$\Gamma_1\oplus\Gamma_2\oplus\Gamma_3$	$\Gamma_4\oplus\Gamma_5$	$\Gamma_4\oplus\Gamma_5$	Γ_8	Γ_8	$\Gamma_6\oplus\Gamma_7\oplus\Gamma_8$
Γ_4				$\Gamma_1\oplus\Gamma_2\oplus\Gamma_4\oplus\Gamma_5$	$\Gamma_2\oplus\Gamma_3\oplus\Gamma_4\oplus\Gamma_5$	$\Gamma_6\oplus\Gamma_8$	$\Gamma_7\oplus\Gamma_8$	$\Gamma_6\oplus\Gamma_7\oplus2\Gamma_8$
Γ_5					$\Gamma_1\oplus\Gamma_3\oplus\Gamma_4\oplus\Gamma_5$	$\Gamma_7\oplus\Gamma_8$	$\Gamma_6\oplus\Gamma_8$	$\Gamma_6\oplus\Gamma_7\oplus2\Gamma_8$
Γ_6						$\Gamma_1\oplus\Gamma_4$	$\Gamma_2\oplus\Gamma_5$	$\Gamma_3\oplus\Gamma_4\oplus\Gamma_5$
Γ_7							$\Gamma_1\oplus\Gamma_4$	$\Gamma_3\oplus\Gamma_4\oplus\Gamma_5$
Γ_8								$\Gamma_1\oplus\Gamma_2\oplus\Gamma_3\oplus2\Gamma_4\oplus2\Gamma_5$

Table 2—continued

(c) Compatibilities with representations of the three-dimensional rotation group

$\mathscr{D}^0$	Γ_1
$\mathscr{D}^1$	Γ_4
$\mathscr{D}^2$	$\Gamma_3 \oplus \Gamma_5$
$\mathscr{D}^3$	$\Gamma_2 \oplus \Gamma_4 \oplus \Gamma_5$
$\mathscr{D}^4$	$\Gamma_1 \oplus \Gamma_3 \oplus \Gamma_4 \oplus \Gamma_5$
$\mathscr{D}^5$	$\Gamma_3 \oplus 2\Gamma_4 \oplus \Gamma_5$
$\mathscr{D}^6$	$\Gamma_1 \oplus \Gamma_2 \oplus \Gamma_3 \oplus \Gamma_4 \oplus 2\Gamma_5$
$\mathscr{D}^7$	$\Gamma_2 \oplus \Gamma_3 \oplus 2\Gamma_4 \oplus 2\Gamma_5$
$\mathscr{D}^8$	$\Gamma_1 \oplus 2\Gamma_3 \oplus 2\Gamma_4 \oplus 2\Gamma_5$
$\mathscr{D}^9$	$\Gamma_1 \oplus \Gamma_2 \oplus \Gamma_3 \oplus 3\Gamma_4 \oplus 2\Gamma_5$
$\mathscr{D}^{10}$	$\Gamma_1 \oplus \Gamma_2 \oplus 2\Gamma_3 \oplus 2\Gamma_4 \oplus 3\Gamma_5$
$\mathscr{D}^{11}$	$\Gamma_2 \oplus 2\Gamma_3 \oplus 3\Gamma_4 \oplus 3\Gamma_5$
$\mathscr{D}^{1/2}$	Γ_6
$\mathscr{D}^{3/2}$	Γ_8
$\mathscr{D}^{5/2}$	$\Gamma_7 \oplus \Gamma_8$
$\mathscr{D}^{7/2}$	$\Gamma_6 \oplus \Gamma_7 \oplus \Gamma_8$
$\mathscr{D}^{9/2}$	$\Gamma_6 \oplus 2\Gamma_8$
$\mathscr{D}^{11/2}$	$\Gamma_6 \oplus \Gamma_7 \oplus 2\Gamma_8$
$\mathscr{D}^{13/2}$	$\Gamma_6 \oplus 2\Gamma_7 \oplus 2\Gamma_8$
$\mathscr{D}^{15/2}$	$\Gamma_6 \oplus \Gamma_7 \oplus 3\Gamma_8$
$\mathscr{D}^{17/2}$	$2\Gamma_6 \oplus \Gamma_7 \oplus 3\Gamma_8$
$\mathscr{D}^{19/2}$	$2\Gamma_6 \oplus 2\Gamma_7 \oplus 3\Gamma_8$
$\mathscr{D}^{21/2}$	$\Gamma_6 \oplus 2\Gamma_7 \oplus 4\Gamma_8$
$\mathscr{D}^{23/2}$	$2\Gamma_6 \oplus 2\Gamma_7 \oplus 4\Gamma_8$

For those values of l and j not tabulated explicitly the reductions of $\mathscr{D}^l$ ($l =$ integer) and $\mathscr{D}^j$ ($j =$ half-odd-integer) are given by

$$\mathscr{D}^{12m+\lambda} = m \text{ reg} \oplus \mathscr{D}^\lambda$$

where the symbol 'reg' is used to denote the single-valued or double-valued irreducible representations, as appropriate, that appear in the regular representation of the point group.

Table 3.   Coupling coefficients for $\Gamma_3 \boxtimes \Gamma_5$ of 432 (O)

	$u_1{}^3 v_1{}^5$	$u_1{}^3 v_2{}^5$	$u_1{}^3 v_3{}^5$	$u_2{}^3 v_1{}^5$	$u_2{}^3 v_2{}^5$	$u_2{}^3 v_3{}^5$
$\psi_1{}^4$	$-\frac{1}{2}\sqrt{3}$	0	0	$-\frac{1}{2}$	0	0
$\psi_2{}^4$	0	$\frac{1}{2}\sqrt{3}$	0	0	$-\frac{1}{2}$	0
$\psi_3{}^4$	0	0	0	0	0	1
$\psi_1{}^5$	$-\frac{1}{2}$	0	0	$\frac{1}{2}\sqrt{3}$	0	0
$\psi_2{}^5$	0	$-\frac{1}{2}$	0	0	$-\frac{1}{2}\sqrt{3}$	0
$\psi_3{}^5$	0	0	1	0	0	0

(Koster *et al.*, 1963)

The idea of a Kronecker product can be extended to the irreducible co-representations of a black and white, or magnetic, point group (Bradley and Davies 1968). Suppose that $D\Gamma_i$, $D\Gamma_j$ and $D\Gamma_k$ are the irreducible co-representations of

$$\mathbf{M} = \mathbf{H} + A\,\mathbf{H} \tag{1.11}$$

derived from the irreducible representations Γ_i, Γ_j and Γ_k, respectively, of $\mathbf{H}$. We then seek to determine the Clebsch–Gordan coefficients $d_{ij,\,k}$ in the reduction

$$D\Gamma_i \boxtimes D\Gamma_j = \sum_k d_{ij,\,k} D\Gamma_k. \tag{1.22}$$

We assume that the Clebsch–Gordan coefficients $c_{ij,\,k}$ in the related reduction

$$\Gamma_i \boxtimes \Gamma_j = \sum_k c_{ij,\,k} \Gamma_k \tag{1.18}$$

for the unitary subgroup $\mathbf{H}$ have already been determined. It is then possible to show that there is quite a simple relationship between $d_{ij,\,k}$ and $c_{ij,\,k}$; the details of this relationship vary, depending on the case ((a), (b) or (c)) to which $D\Gamma_i$, $D\Gamma_j$ and $D\Gamma_k$ belong (see table 5 of Bradley and Davies (1968) or table 7.8 of Bradley and Cracknell (1972)). An example of the reduction of $D\Gamma_i \boxtimes D\Gamma_j$ for the point group $4'/mmm'$ will be encountered later (see table 17).

A special situation arises if we consider the Kronecker product $\Gamma_i \boxtimes \Gamma_j$ when $i = j$, that is when we consider the product of a representation, Γ_i, with itself. This product, $\Gamma_i \boxtimes \Gamma_i$, is referred to as the square of Γ_i. Suppose that

$$|u_p{}^i\rangle = |u_1{}^i,\, u_2{}^i,\, \dots,\, u_d{}^i\rangle \tag{1.23}$$

is a basis of Γ_i, where d is the dimension of Γ_i. Suppose that $d > 1$. If we consider a vector in which each component is an ordered pair of functions $(u_p{}^i, u_q{}^i)$, where $1 \leqslant p \leqslant d$ and $1 \leqslant q \leqslant d$, this will form a basis of $\Gamma_i \boxtimes \Gamma_i$. It is possible to find some unitary transformation so that each component of the basis of $\Gamma_i \boxtimes \Gamma_i$ is either a symmetrized combination of the form $(u_p{}^i, u_q{}^i) + (u_q{}^i, u_p{}^i)$ or an anti-symmetrized combination of the form $(u_p{}^i, u_q{}^i) - (u_q{}^i, u_p{}^i)$. The square $\Gamma_i \boxtimes \Gamma_i$ can therefore be separated into two parts, the symmetrized square, which is often denoted by $[\Gamma_i \boxtimes \Gamma_i]$, to which the $\frac{1}{2}d(d+1)$ functions $(u_p{}^i, u_q{}^i) + (u_q{}^i, u_p{}^i)$ belong and the anti-symmetrized square, which is often denoted by $\{\Gamma_i \boxtimes \Gamma_i\}$, to which the $\frac{1}{2}d(d-1)$ functions $(u_p{}^i, u_q{}^i) - (u_q{}^i, u_p{}^i)$ belong. The identification of $[\Gamma_i \boxtimes \Gamma_i]$ and $\{\Gamma_i \boxtimes \Gamma_i\}$ for each of the 32 classical point groups is given, for example, in § 2.6 of Bradley and Cracknell (1972). The idea of symmetrized and anti-symmetrized squares can readily be extended to higher powers such as symmetrized and anti-symmetrized cubes, etc.; the formal theory is given, for example, in chapter 4 of the book by Lyubarskii (1960). In practice the symmetrized cubes have been found to be useful, see § 8.4, but no use has been found so far for other symmetrized or anti-symmetrized products of order greater than two. Tables of the symmetrized cubes for the classical point groups have been given by Cracknell and Joshua (1968). The ideas of symmetrized and anti-symmetrized products can also be extended to the Kronecker products of co-representations of black and white point groups. Tables of the anti-symmetrized squares and the symmetrized cubes of the irreducible co-representations of each of the black and white, or magnetic, point groups have been given by Cracknell and Sedaghat (1972).

One situation in which symmetrized and anti-symmetrized products can be expected to be important arises when one considers the wave function for a collection of identical particles by taking products of one-particle wave functions. Another example occurs in connection with the theory of continuous, or second-order, phase transitions (see § 8.4).

§ 2. THE EARLY APPLICATIONS

In this section we shall remind the reader about two of the very early applications of group theory which are now quite familiar to most solid-state physicists. The first involves localized electronic states and uses point-group representations, while the second involves itinerant electrons and uses space-group representations. As indicated already in the introduction, we have separated these two well-established applications so that we can discuss them quite briefly ; for further details we refer the reader to the several established textbooks and review articles mentioned in § 1.1. Recent developments of these two applications will be described in some detail in § 4 and in § 6.

2.1. *Crystal-field theory*

The earliest example of the application of group theory in solid-state physics was in connection with the splitting of atomic energy levels in crystalline solids (Bethe 1929). In § 1.3 we have already noted that the eigenvalues and eigenfunctions of an electron must belong to the irreducible representations of the group of the symmetry operations of the Hamiltonian, $\mathscr{H}$, of the system in which the electron finds itself (Wigner's theorem). In the case of a free atom the appropriate group of symmetry operations is the three-dimensional rotation group. This is reflected in the established scheme which is used in the labelling of the energy levels of electrons in atoms, particularly the angular momentum quantum number, l, and the magnetic quantum number, m_l. Ordinary crystal-field theory, as initiated by Bethe (1929), is concerned with studying the degeneracies of the energy levels of electrons that are localized on atoms which are situated in a crystalline solid. If an atom is part of a crystal structure the electrostatic potential which is experienced by an electron in that atom will no longer be spherically symmetrical but will have the symmetry of one of the crystallographic point groups, **G**. Therefore the wave function of the electron must belong to one of the irreducible representations of the point group **G**, instead of to one of the irreducible representations of the rotation group. The degeneracies of the energy levels of the electrons will be determined by the degeneracies of the irreducible representations of **G**. This reduction in the symmetry of the Hamiltonian, because of the surrounding atoms in the crystal, will therefore in general lead to a reduction in the degeneracy of the energy levels of the electrons, relative to the degeneracies of the energy levels of the electrons in free atoms of the same element. The details of the group-theoretical analysis of these degeneracies, neglecting spin and following the method described in detail in the paper by Bethe (1929) will be illustrated shortly by an example.

The extension of the work of Bethe to include the effect of electron spin was due to Opechowski (1940). When electron spin is included there are various possibilities that have to be considered, depending on the relative strength of the crystal field and the spin–orbit coupling. They are :

(i) free atom neglecting spin–orbit coupling,
(ii) free atom with spin–orbit coupling,
(iii) strong spin–orbit coupling and weak crystal field,
(iv) weak spin–orbit coupling and strong crystal field,
(v) crystal field (no spin–orbit coupling).

The procedure for determining the splitting of any given term for various relative strengths of spin–orbit coupling and of the crystalline field is now quite well established (see, for example, chapter 4 of the book by Tinkham (1964)). Cases (i) and (ii) are not confined to crystal-field theory but belong to the general theory of atomic spectra. Case (v) is the situation treated by Bethe (1929) in which electron spin was ignored. The sequence in which crystal-field theory proceeds is usually either from (i) to (v) and then back to (iv) or from (i) to (ii) and then on to (iii). Each of these two sequences involves two steps, one is the subduction of an irreducible representation of one group onto one of its subgroups and the other is the reduction of the inner Kronecker product of two representations of a given group. The inner Kronecker product will either involve two single-valued representations if S is an integer, or one single-valued and one double-valued representation if S is half an odd integer. In the first sequence the subduction is performed first :

(A) (i)→(v), the subduction of the irreducible representation $\mathscr{D}^L$ of $O(3)$, the three-dimensional rotation group, on to the point group **G**.
(B) (v)→(iv), the reduction of the Kronecker product of the irreducible representation $\mathscr{D}^S$ (of $O(3)$, subduced onto **G**) and the irreducible representations of **G** obtained in (A).

In the second sequence the reduction of the Kronecker product is performed first :

(C) (i)→(ii), the reduction of the Kronecker product of $\mathscr{D}^L$ and $\mathscr{D}^S$ of $O(3)$, corresponding to the orbital and spin angular momenta, to give the irreducible representations $\mathscr{D}^J$ of $O(3)$.
(D) (ii)→(iii), the subduction of $\mathscr{D}^J$ of $O(3)$ on to the point group **G** (these irreducible representations will be single-valued or double-valued depending on whether the value of J is an integer or half an odd integer).

The various subductions, either from $O(3)$ to a point group **G**, or from a point group $\mathbf{G}_1$ to one of its subgroups $\mathbf{G}_2$, can all be performed very easily using the compatibility tables given by Koster *et al.* (1963). The continuity of cases (iii) and (iv), from strong spin–orbit coupling and weak crystal field to weak spin–orbit coupling and strong crystal field is established by inspection of the two cases and making them fit together as the relative strengths of the spin–orbit coupling and the crystal field are varied.

We illustrate this by considering, as an example, the splitting of a ^4F term in a crystal field that possesses the symmetry of the cubic point group 432 (O). The spin degeneracy of this term is 4 and the orbital degeneracy is $(2L+1)=7$, so that the total degeneracy is 28. We consider the four steps

Fig. 7

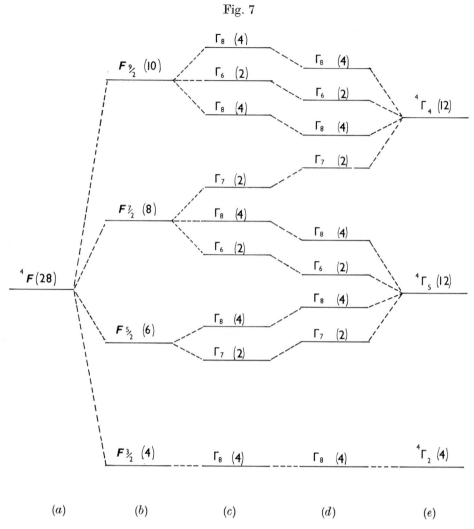

The splitting of an atomic ^4F term in 432 (O) ; (a) free atom, (b) free atom with spin–orbit coupling, (c) strong spin–orbit coupling and weak crystal field, (d) weak spin–orbit coupling and strong crystal field, and (e) crystal field alone.

(A), (B), (C) and (D) outlined above. In (A) the free atom with no spin–orbit coupling is placed in a situation with a strong crystal field belonging to 432 (O). The subduction of $\mathscr{D}^3$ of $O(3)$ on to the point group 432 (O) gives

$$\mathscr{D}^3 = \Gamma_2 \oplus \Gamma_4 \oplus \Gamma_5 \qquad (2.1)$$
$$(7) \quad (1) \quad (3) \quad (3)$$

where the numbers in brackets indicate the degeneracy ; so far spin is ignored. This splitting is illustrated in fig. 7. In (B) the addition of weak spin–orbit coupling to the crystal field is described by taking the Kronecker product of $\mathscr{D}^S$ with each of the irreducible representations of 432 (O) which were found in (v). Since $S = \frac{3}{2}$, $\mathscr{D}^S$ when restricted to 432 (O) is Γ_8, see table 2. That is, Γ_2 splits according to

$$\Gamma_8 \boxtimes \Gamma_2 = \Gamma_8 \qquad (2.2)$$
$$(4) \quad (1) \quad (4)$$

where we have used table 2. Similarly, Γ_4 and Γ_5 split according to

$$\Gamma_8 \boxtimes \Gamma_4 = \Gamma_6 \oplus \Gamma_7 \oplus 2\Gamma_8 \qquad (2.3)$$
$$(4) \quad (3) \quad (2) \quad (2) \quad 2(4)$$

and

$$\Gamma_8 \boxtimes \Gamma_5 = \Gamma_6 \oplus \Gamma_7 \oplus 2\Gamma_8. \qquad (2.4)$$
$$(4) \quad (3) \quad (2) \quad (2) \quad 2(4)$$

The introduction of spin–orbit coupling therefore splits each of the levels in (v) with three-fold orbital degeneracy into two two-fold degenerate levels and two four-fold degenerate levels, see fig. 7. Now step (C), which covers the case of adding spin–orbit coupling to the free atom, produces a splitting determined by $\mathscr{D}^J = \mathscr{D}^L \boxtimes \mathscr{D}^S$, that is

$$\mathscr{D}^3 \boxtimes \mathscr{D}^{3/2} = \mathscr{D}^{9/2} \oplus \mathscr{D}^{7/2} \oplus \mathscr{D}^{5/2} \oplus \mathscr{D}^{3/2} \qquad (2.5)$$
$$(7) \quad (4) \quad (10) \quad (8) \quad (6) \quad (4)$$

that is, all the representations with $J = L + S$, $L + S - 1$, ..., $|L - S|$ appear exactly once each, see fig. 7. (D) describes the addition of a weak crystal field with the symmetry of 432 (O) to the situation of strong spin–orbit coupling ; this is determined by the subduction of $\mathscr{D}^J$ onto the point group 432 (O) and we find, using table 2, that

$$\mathscr{D}^{3/2} = \Gamma_8 \qquad (2.6)$$
$$(4) \quad (4)$$

$$\mathscr{D}^{5/2} = \Gamma_7 \oplus \Gamma_8 \qquad (2.7)$$
$$(6) \quad (2) \quad (4)$$

$$\mathscr{D}^{7/2} = \Gamma_6 \oplus \Gamma_7 \oplus \Gamma_8 \qquad (2.8)$$
$$(8) \quad (2) \quad (2) \quad (4)$$

and

$$\mathscr{D}^{9/2} = \Gamma_6 \oplus 2\Gamma_8. \qquad (2.9)$$
$$(10) \quad (2) \quad 2(4)$$

These splittings are illustrated in fig. 7. There must be a continuous variation from (iii) to (iv) as the relative strengths of the crystal field and the spin–orbit interaction are varied.

The early work on crystal-field theory which we have described in this section was primarily concerned with studying the qualitative features of the splitting patterns of atomic energy levels in crystalline solids. The importance of this theory was in helping to interpret spectroscopic measurements on localized electronic states in crystalline solids. We use the term ' spectroscopic measurements ' to include not only the results of optical work but also measurements involving radiation in various other regions of the electromagnetic spectrum, as well as the results of various types of magnetic resonance experiments. Various later developments will be described in some of the subsequent sections. These include the study of the wave functions themselves and attempts to obtain quantitative estimates of the splittings (§§ 4.1, 4.2 and 4.4) and also the effect of magnetic ordering on the splitting of atomic energy levels (§ 4.5).

2.2. *The labelling and degeneracy of electronic energy bands*

The other early application of group theory in solid-state physics was in connection with the labelling and degeneracies of electronic energy bands in crystals and has its origins in the work of Bouckaert *et al.* (1936). In this case, since we are dealing with itinerant electrons instead of localized electrons, the appropriate Hamiltonian will possess the symmetry of one of the space groups. Therefore, making use of Wigner's theorem (see § 1.3) the wave functions and energy eigenvalues must belong to the irreducible representations of a space group rather than of a point group. That is, the wave vector **k** forms a good quantum number for the labelling of the electronic energy bands of a crystalline solid. In practice the crystals considered in the early days were nearly always metals.

As an illustration we consider the case of the much-studied space group $Fm3m$ (O_h^5) which describes a number of structures including that of a face-centred cubic metal. The Brillouin zone for this space group is illustrated in fig. 8 (*a*) while the character tables for each of the special points of symmetry

Fig. 8

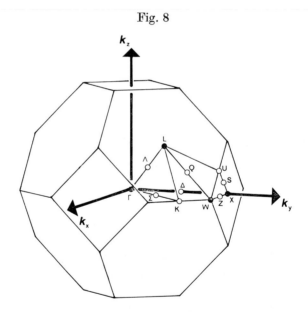

Brillouin zone for the space group $Fm3m$ (O_h^5).

and lines of symmetry are given in table 4. The wave function of an electron with any given wave vector must belong to one of those irreducible representations of the space group $Fm3m$ (O_h^5) corresponding to that wave vector, **k**. For an arbitrary wave vector in the Brillouin zone there is only one (non-degenerate) irreducible representation and so one would expect each energy band to be non-degenerate there. At a wave vector corresponding to a

Table 4. Character tables for the irreducible representations of $Fm3m$ (O_h^5)

Γ	E	$8C_{3j}^{\pm}$	$3C_{2m}$	$6C_{4m}^{\pm}$	$6C_{2p}$
	I	$8S_{6j}^{\mp}$	$3\sigma_m$	$6S_{4m}^{\mp}$	$6\sigma_{dp}$
$m3m$ (O_h)					
$\Gamma_1,\ \Gamma_1'$	1	1	1	1	1
$\Gamma_2,\ \Gamma_2'$	1	1	1	-1	-1
$\Gamma_{12},\ \Gamma_{12}'$	2	-1	2	0	0
$\Gamma_{15}',\ \Gamma_{15}$	3	0	-1	1	-1
$\Gamma_{25}',\ \Gamma_{25}$	3	0	-1	-1	1

$\Sigma,\ K$			E	C_{2a}	σ_z	σ_{db}
	$S,\ U$		E	C_{2c}	σ_{de}	σ_y
		Z	E	C_{2x}	σ_z	σ_y
$mm2$ (C_{2v})						
$\Sigma_1,\ K_1$	$S_1,\ U_1$	Z_1	1	1	1	1
$\Sigma_2,\ K_2$	$S_2,\ U_2$	Z_2	1	1	-1	-1
$\Sigma_4,\ K_4$	$S_3,\ U_3$	Z_3	1	-1	1	-1
$\Sigma_3,\ K_3$	$S_4,\ U_4$	Z_4	1	-1	-1	1

Δ			E	C_{2y}	$C_{4y}^{\pm}$	$\sigma_z,\ \sigma_x$	$\sigma_{de},\ \sigma_{dc}$
	X		$\{E$	C_{2y}	$C_{4y}^{\mp}$	$C_{2x},\ C_{2z}$	$C_{2e},\ C_{2c}$
			$\{I$	σ_y	$S_{4y}^{\pm}$	$\sigma_x,\ \sigma_z$	$\sigma_{de},\ \sigma_{dc}$
		W	E	C_{2x}	$S_{4x}^{\pm}$	$C_{2d},\ C_{2f}$	$\sigma_y,\ \sigma_z$
$4mm$ (C_{4v})	**$4/mmm$ (D_{4h})**	**$\bar{4}2m$ (D_{2d})**					
Δ_1	$X_1,\ X_1'$	W_1	1	1	1	1	1
Δ_1'	$X_4,\ X_4'$	W_2	1	1	1	-1	-1
Δ_2	$X_2,\ X_2'$	W_1'	1	1	-1	1	-1
Δ_2'	$X_3,\ X_3'$	W_2'	1	1	-1	-1	1
Δ_5	$X_5,\ X_5'$	W_3	2	-2	0	0	0

Λ		E	$C_{31}^{\pm}$	$\sigma_{db},\ \sigma_{de},\ \sigma_{df}$
	L	$\{E$	$C_{31}^{\pm}$	$C_{2b},\ C_{2e},\ C_{2f}$
		$\{I$	$S_{61}^{\mp}$	$\sigma_{db},\ \sigma_{de},\ \sigma_{df}$
$3m$ (C_{3v})	**$\bar{3}m$ (D_{3d})**			
Λ_1	$L_1,\ L_1'$	1	1	1
Λ_2	$L_2,\ L_2'$	1	1	-1
Λ_3	$L_3,\ L_3'$	2	-1	0

Table 4—continued

C		E	σ_{db}
O		E	σ_z
J		E	σ_{de}
B		E	σ_y
	Q	E	C_{2f}
$m(C_{1h})$	$2(C_2)$		
$C^+O^+J^+B^+$	Q^+	1	1
$C^-O^-J^-B^-$	Q^-	1	-1

Notes. (i). The symmetry operations are labelled in the notation of Bradley and Cracknell (1972) and the irreducible representations are labelled in the notation of Bouckaert *et al.* (1936). (ii). For direct product groups we only give one quarter of the character table. If R is an element for which $\chi(R)$ is given in the table, then $\chi(IR) = +\chi(R)$, where I is the inversion, for representations in the first column for each **k** (such as Γ_{12} for example) and $\chi(IR) = -\chi(R)$ for representations given in the second column (such as Γ_{12}' for example).

special point or line of symmetry the degeneracies of the energy bands will be determined by the degeneracies of the irreducible representations for that wave vector. Thus, for example, at Γ the bands may be non-degenerate, two-fold degenerate, or three-fold degenerate. The degeneracy is given by the character of the identity element, E, at Γ in table 4 ; thus we have

$$\text{non-degenerate} : \Gamma_1, \Gamma_1', \Gamma_2, \Gamma_2',$$

$$\text{two-fold degenerate} : \Gamma_{12}, \Gamma_{12}',$$

$$\text{three-fold degenerate} : \Gamma_{15}, \Gamma_{15}', \Gamma_{25}, \Gamma_{25}'.$$

Similar arguments apply to the degeneracies of the bands at the other special points of symmetry in the Brillouin zone and also along the lines of symmetry ; thus, for example, all the bands are non-degenerate at Z or along Σ. Along Δ the bands are either non-degenerate, $\Delta_1, \Delta_1', \Delta_2$ and Δ_2', or two-fold degenerate, Δ_5.

The connectivities of the bands between points of symmetry, lines of symmetry, and general wave vectors can be established by studying the compatibilities between the irreducible representations for the various wave vectors. Suppose that $\mathbf{k}_0$ is a wave vector for a point of symmetry and $(\mathbf{k}_0 + \boldsymbol{\varkappa})$ is a wave vector for a line of symmetry passing through $\mathbf{k}_0$. Then $\mathbf{G}^{\mathbf{k}_0+\boldsymbol{\varkappa}}$ is a subgroup of $\mathbf{G}^{\mathbf{k}_0}$ and the connectivities between the bands at $\mathbf{k}_0$ and $\mathbf{k}_0 + \boldsymbol{\varkappa}$ can be established by subduction of the representations $\Gamma_i^{\mathbf{k}_0}$ of the supergroup $\mathbf{G}^{\mathbf{k}_0}$ on to the subgroup $\mathbf{G}^{\mathbf{k}_0+\boldsymbol{\varkappa}}$. For a symmorphic space group this means that we have to study the relationship between the representations of the point group $\mathbf{P}(\mathbf{k}_0)$ and of its subgroup $\mathbf{P}(\mathbf{k}_0 + \boldsymbol{\varkappa})$. Compatibility tables for the various points, lines, and planes of symmetry for the space group $Fm3m$ $(O_h{}^5)$ are given in table 5. With the aid of these compatibility tables it is possible to study the connectivities of the energy bands for a material with a structure belonging to this space group. Thus, for example, along Δ a three-fold degenerate band that belongs to Γ_{15} at Γ splits into two

Table 5. Compatibility relations for the f.c.c. structure

(i) Γ : Δ, Λ, Σ				
Γ_1	Γ_2	Γ_{12}	$\Gamma_{15}{}'$	$\Gamma_{25}{}'$
Δ_1	Δ_2	$\Delta_1\Delta_2$	$\Delta_1{}'\Delta_5$	$\Delta_2{}'\Delta_5$
Λ_1	Λ_2	Λ_3	$\Lambda_2\Lambda_3$	$\Lambda_1\Lambda_3$
Σ_1	Σ_4	$\Sigma_1\Sigma_4$	$\Sigma_2\Sigma_3\Sigma_4$	$\Sigma_1\Sigma_2\Sigma_3$
$\Gamma_1{}'$	$\Gamma_2{}'$	$\Gamma_{12}{}'$	Γ_{15}	Γ_{25}
$\Delta_1{}'$	$\Delta_2{}'$	$\Delta_1{}'\Delta_2{}'$	$\Delta_1\Delta_5$	$\Delta_2\Delta_5$
Λ_2	Λ_1	Λ_3	$\Lambda_1\Lambda_3$	$\Lambda_2\Lambda_3$
Σ_2	Σ_3	$\Sigma_2\Sigma_3$	$\Sigma_1\Sigma_3\Sigma_4$	$\Sigma_1\Sigma_2\Sigma_4$

(ii) X : Δ, Z, S									
X_1	X_2	X_3	X_4	X_5	$X_1{}'$	$X_2{}'$	$X_3{}'$	$X_4{}'$	$X_5{}'$
Δ_1	Δ_2	$\Delta_2{}'$	$\Delta_1{}'$	Δ_5	$\Delta_1{}'$	$\Delta_2{}'$	Δ_2	Δ_1	Δ_5
Z_1	Z_1	Z_4	Z_4	Z_2Z_3	Z_2	Z_2	Z_3	Z_3	Z_1Z_4
S_1	S_4	S_1	S_4	S_2S_3	S_2	S_3	S_2	S_3	S_1S_4

(iii) L : Λ, Q					
L_1	L_2	L_3	$L_1{}'$	$L_2{}'$	$L_3{}'$
Λ_1	Λ_2	Λ_3	Λ_2	Λ_1	Λ_3
Q^+	Q^-	Q^+Q^-	Q^+	Q^-	Q^+Q^-

(iv) W : Z, Q				
W_1	W_2	$W_1{}'$	$W_2{}'$	W_3
Z_1	Z_2	Z_2	Z_1	Z_3Z_4
Q^+	Q^-	Q^+	Q^-	Q^+Q^-

(v) Symmetry planes		
Plane	$+$	$-$
O (ΓWX)	Σ_1, Σ_4	Σ_2, Σ_3
	$\Delta_1, \Delta_2, \Delta_5$	$\Delta_1{}', \Delta_2{}', \Delta_5$
	Z_1, Z_3	Z_2, Z_4
C (ΓLK)	Σ_1, Σ_3	Σ_2, Σ_4
	Λ_1, Λ_3	Λ_2, Λ_3
J (ΓXL)	Λ_1, Λ_3	Λ_2, Λ_3
	S_1, S_3	S_2, S_4
	$\Delta_1, \Delta_2{}', \Delta_5$	$\Delta_2, \Delta_1{}', \Delta_5$
B (XUW)	S_1, S_4	S_2, S_3
	Z_1, Z_4	Z_2, Z_3

bands Δ_1 (non-degenerate) and Δ_5 (two-fold degenerate), but along Σ the Γ_{15} band splits into three non-degenerate bands (Σ_1, Σ_3 and Σ_4). As an example to illustrate the connectivities and the use of the compatibility tables, we include the band structure of metallic sodium, calculated by Kenney, in fig. 9. In this case, although we are dealing with itinerant electrons, it is still possible to discern the origins of these bands in terms of the energy levels of free sodium atoms ; we shall return to this aspect in § 6.1.

If all that could be achieved by the use of group theory in connection with electronic band structures were a scheme for labelling the energy bands and understanding their degeneracies, it is unlikely that group theory would ever have achieved the important position which it now occupies in this subject. Until the quite recent introduction of sophisticated many-body techniques, band-structure calculations were based on the use of the independent-particle approximation. Methods based on the use of this approximation involve constructing, by some method, a potential $V(\mathbf{r})$ to be used in the one-electron Schrödinger equation

$$-\frac{\hbar^2}{2m}\,\nabla^2\psi_j(\mathbf{k},\,\mathbf{r}) + V(\mathbf{r})\psi_j(\mathbf{k},\,\mathbf{r}) = E_j(\mathbf{k})\psi_j(\mathbf{k},\,\mathbf{r}) \qquad (2.10)$$

and then choosing one of the several available methods for solving this equation subject to the appropriate boundary conditions for a specimen of crystalline solid. About 15 years ago, when reliable quantitative results were beginning to be obtained from band-structure calculations for many simple crystalline solids, group theory was an essential ingredient in such calculations. This was because group theory enables the calculations to be simplified quite substantially for wave vectors $\mathbf{k}$ of special symmetry in the Brillouin zone, while calculations for general wave vectors $\mathbf{k}$ were outside the scope of the computing machinery then available.

Suppose that, for a given representation $\Gamma_j^{\mathbf{k}}$, the wave function $\psi_j(\mathbf{k},\,\mathbf{r})$, or rather $u_j(\mathbf{k},\,\mathbf{r})$, is expanded in terms of spherical harmonics or plane waves, giving

$$u_j(\mathbf{k},\,\mathbf{r}) = \sum_{n,\,l,\,m} A_{nlm}{}^{j\mathbf{k}} R_{nl}(r)\, Y_l{}^m(\theta,\,\phi) \qquad (2.11)$$

or

$$u_j(\mathbf{k},\,\mathbf{r}) = \sum_{n_1,\,n_2,\,n_3} C_{n_1 n_2 n_3}{}^{j\mathbf{k}} \exp{(i\mathbf{G}_{n_1 n_2 n_3} \cdot \mathbf{r})} \qquad (2.12)$$

respectively. $\mathbf{G}_{n_1 n_2 n_3}$ is a reciprocal lattice vector given by

$$\mathbf{G}_{n_1 n_2 n_3} = n_1\mathbf{g}_1 + n_2\mathbf{g}_2 + n_3\mathbf{g}_3 \qquad (2.13)$$

where n_1, n_2 and n_3 are integers. In band-structure calculations the expansion in eqn. (2.11) is used all over the Wigner–Seitz unit cell in the cellular method and in the inner regions of the unit cell in the augmented plane wave (A.P.W.) method. The expansion in eqn. (2.12) is used in the orthogonalized plane wave (O.P.W.) method and in the outer regions of the unit cell in the augmented plane wave method. By applying the operator W_{pr} (see eqn. (1.10)) to an expansion of the form given in eqn. (2.11) or (2.12) it is possible to eliminate some of the terms in the expansion and still obtain an expansion that is complete for $\Gamma_j^{\mathbf{k}}$. In other words, at the outset of a calculation $\psi_j(\mathbf{k},\,\mathbf{r})$ is expanded in terms of functions which are symmetry-adapted to

Fig. 9

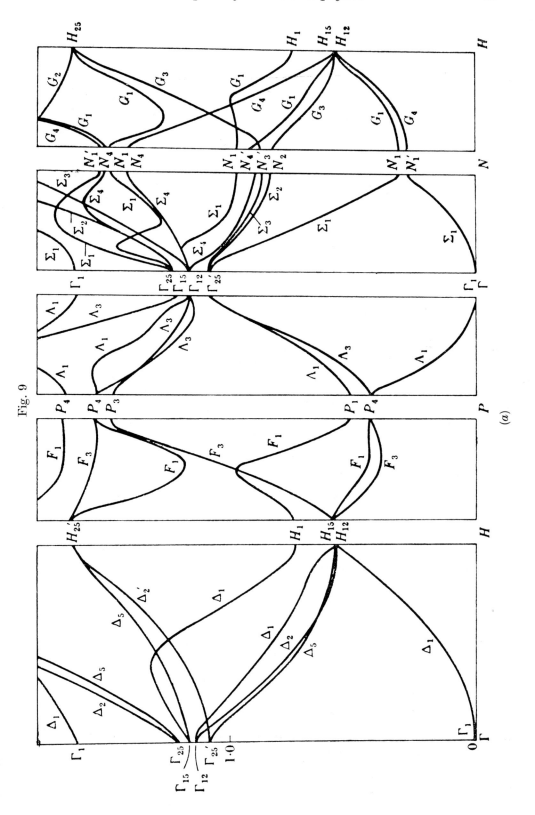

(a)

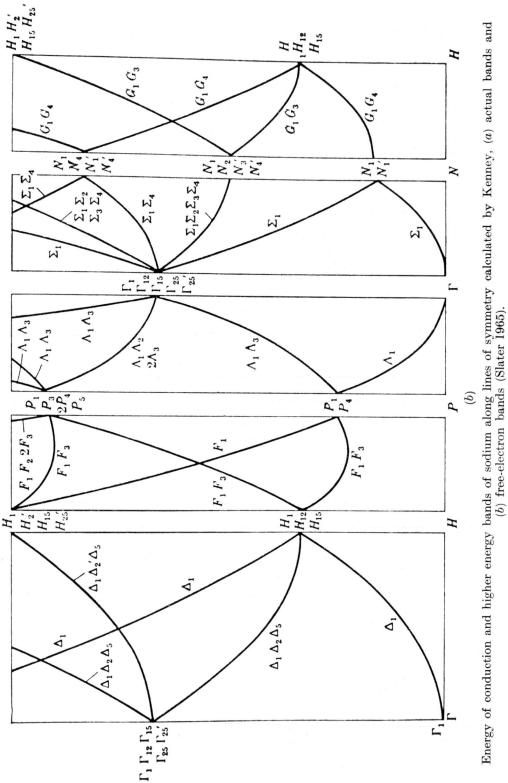

Energy of conduction and higher energy bands of sodium along lines of symmetry calculated by Kenney, (a) actual bands and (b) free-electron bands (Slater 1965).

$\Gamma_j{}^k$. Further details of the determination of symmetry-adapted functions are described for spherical harmonics by Altmann and Bradley (1965) and Altmann and Cracknell (1965) and for plane waves by Luehrmann (1968) and Cornwell (1969). Symmetry-adapted forms of $\psi_j(\mathbf{k}, \mathbf{r})$ for each of the irreducible representations of the group of $\mathbf{k}$, for each special wave vector $\mathbf{k}$, can then be used in turn in solving eqn. (2.10) ; in this manner the group-theoretical labels are automatically assigned correctly to the various calculated energy bands.

§ 3. Tensor properties

In this part of the article we shall be concerned with the application of Neumann's principle in connection with the macroscopic properties of a crystalline material. Many of the quantities that are handled in physics can be adequately described by a scalar, that is by a single number, possibly with a + or − sign attached. However, there are many physical properties of crystalline materials which cannot be described by scalars but which must be described by tensors of various rank. For example, one commonly regards the electric susceptibility of a material as being specified by a single number, but this is only the case for a cubic crystalline material or for an isotropic non-crystalline material such as a glass or polymer. For a crystalline solid which belongs to any one of the crystal systems other than the cubic system the electric susceptibility cannot be described completely by a single number ; instead it has to be described by a tensor of rank 2. The relationship between the electric polarization, $\mathbf{P}$, and the electric field, $\mathbf{E}$, takes the form

$$P_i = \epsilon_0 \sum_j \chi_{ij}{}^e E_j \tag{3.1}$$

where the vectors $\mathbf{P}$ $(= (P_1, P_2, P_3))$ and $\mathbf{E}$ $(= (E_1, E_2, E_3))$ are referred to a set of orthogonal axes $Ox_1x_2x_3$ (or $Oxyz$). There are many other physical properties of crystalline materials which, like the electric susceptibility, cannot be described by a scalar but have to be described by a tensor of rank 1, 2, 3 or even 4.

3.1. *Equilibrium properties of non-magnetic crystals*

Suppose that the symmetry of a given crystal is described by a certain point group $\mathbf{G}$ and that R is a symmetry operation of $\mathbf{G}$. The effect of the point-group operation R on a vector $\mathbf{x}$ $(= (x_1, x_2, x_3))$ is to generate a vector $\mathbf{x}'$ $(= (x_1', x_2', x_3'))$; for each R we can define an orthogonal matrix R_{ij} so that

$$x_1' = \sum_p R_{ip} x_p. \tag{3.2}$$

A vector is a special case of a tensor, namely a tensor of rank 1, and eqn. (3.2) is a special case of the transformation equation which can be regarded as defining a tensor. In general a tensor $d_{ijk \dots n}$ of arbitrary rank is defined by its transformation properties under the rotation R ; indicating the trans-formed tensor by a prime we have

$$d'_{ijk \dots n} = \sum_{p, q, r, \dots, u} R_{ip} R_{jq} R_{kr} \cdots R_{nu} d_{pqr \dots u}. \tag{3.3}$$

An extensive discussion of this defining relation is given, for example, in the first two chapters of the book by Nye (1957). Strictly speaking it is necessary to distinguish between *polar tensors* and *axial tensors*. Equation (3.3) only applies to a polar tensor and has to be replaced by

$$d'_{ijk \ldots n} = |\mathbf{R}| \sum_{p, q, r, \ldots, u} R_{ip} R_{jq} R_{kr} \ldots R_{nu} d_{pqr \ldots u} \tag{3.4}$$

for an axial tensor. The distinction between polar and axial tensors lies in their behaviour under improper rotations for which the $|\mathbf{R}|$ in eqn. (3.4) is negative.

Equations (3.3) and (3.4) are quite general and they apply to any proper or improper rotation R, irrespective of whether R is a symmetry operation of a crystal or not. Let us suppose that $d_{ijk \ldots n}$ is used to represent a certain physical property of a given crystal. If we then consider only those operations R which are symmetry operations of the crystal it is possible to invoke Neumann's principle, which we have already described in § 1.3. From Neumann's principle we know that since R is a symmetry operation of the crystal it must also be a symmetry operation of any physical property of the crystal. In particular R will be a symmetry operation of $d_{ijk \ldots n}$ so that

$$d'_{ijk \ldots n} = d_{ijk \ldots n}. \tag{3.5}$$

Therefore when R is a symmetry operation of the crystal eqns. (3.3) and (3.4) become

$$d_{ijk \ldots n} = \sum_{p, q, r, \ldots, u} R_{ip} R_{jq} R_{kr} \ldots R_{nu} d_{pqr \ldots u} \tag{3.6}$$

for a polar tensor and

$$d_{ijk \ldots n} = |\mathbf{R}| \sum_{p, q, r, \ldots, u} R_{ip} R_{jq} R_{kr} \ldots R_{nu} d_{pqr \ldots u} \tag{3.7}$$

for an axial tensor.

By using eqn. (3.6) or eqn. (3.7), as appropriate, with the various symmetry operations R of a crystal it may be possible to simplify the form of the tensor $d_{ijk \ldots n}$ for that crystal. That is, it may be possible to show that certain tensor components must vanish or to obtain simple numerical relationships among some of the tensor components. It may be possible to show that for a given point group eqn. (3.6) or eqn. (3.7) leads to the vanishing of *all* the components of a given tensor $d_{ijk \ldots n}$; if this occurs the phenomenon that is represented by $d_{ijk \ldots n}$ will be a 'forbidden' phenomenon in a crystal with the symmetry of that point group. The use of 'forbidden' phenomena in connection with the determination of crystal structures will be discussed in § 8.1. It is not necessary to use eqn. (3.6) or eqn. (3.7) for all the symmetry operations in the point group **G** of a crystal, but only for the generating elements of that point group. The results for polar and axial tensors of rank 0, 1, 2, 3 and 4 have been evaluated for all the 32 classical point groups (Fumi 1952, Fieschi 1957) and they are conveniently listed in table 4 of the book by Birss (1964).

The simplifications that arise in the form of a tensor $d_{ijk \ldots n}$ as a result of the use of eqns. (3.6) and (3.7) can be regarded as being a consequence of crystallographic symmetry and, of course, of the application of Neumann's principle. However, in addition to the symmetry imposed by the crystal structure, a tensor that describes a physical property may also possess some intrinsic symmetry arising from the physical nature of the property itself. For example, for the second-rank polar tensors representing the electric susceptibility, χ_{ij}^{e}, the relative permittivity, ϵ_{ij}^{r}, the magnetic susceptibility, χ_{ij}^{m}, and the relative permeability, μ_{ij}^{r}, it is possible to use arguments based on energy considerations (see, for example, pp. 57–60 and 74 of Nye (1957)) to show that, irrespective of any crystallographic considerations, each of these tensors must be a symmetric tensor. As an example of the results obtained by using the crystallographic symmetry and the intrinsic symmetry of a property tensor we list the form of a symmetric polar tensor of rank two for each of the classical point groups in table 6.

Table 6. The form of a symmetric polar tensor of rank two for each of the classical point groups

Point group	d_{ij}
$1, \bar{1}$	$\begin{pmatrix} d_{11} & d_{12} & d_{13} \\ d_{12} & d_{22} & d_{23} \\ d_{13} & d_{23} & d_{33} \end{pmatrix}$
$2, m, 2/m$	$\begin{pmatrix} d_{11} & d_{12} & 0 \\ d_{12} & d_{22} & 0 \\ 0 & 0 & d_{33} \end{pmatrix}$
$222, mm2, mmm$	$\begin{pmatrix} d_{11} & 0 & 0 \\ 0 & d_{22} & 0 \\ 0 & 0 & d_{33} \end{pmatrix}$
$3, \bar{3}, 32, 3m, \bar{3}m$ $4, \bar{4}, 4/m, 422, 4mm, \bar{4}2m, 4/mmm$ $6, \bar{6}, 6/m, 622, 6mm, \bar{6}2m, 6/mmm$	$\begin{pmatrix} d_{11} & 0 & 0 \\ 0 & d_{11} & 0 \\ 0 & 0 & d_{33} \end{pmatrix}$
$23, m3, 432, \bar{4}3m, m3m$	$\begin{pmatrix} d_{11} & 0 & 0 \\ 0 & d_{11} & 0 \\ 0 & 0 & d_{11} \end{pmatrix}$

In addition to the various second-rank tensors which we have just mentioned there are several other tensors which are used to represent important physical properties of crystalline materials. These too can be simplified by exploiting the appropriate point-group symmetry and comprehensive accounts will be found in the book by Nye (1957) and the review article by Smith (1958). We shall just mention one or two further examples to illustrate that such tensors may be of rank other than 2. Consider ferroelectricity and pyroelectricity. A ferroelectric crystal is a crystal which possesses a spontaneous electric dipole moment **P** $(P_i = (P_1, P_2, P_3))$ in the absence of an external electric field. A pyroelectric crystal is one that either develops an electric dipole moment when heated or, if it is already ferroelectric, changes its dipole moment when heated. The change in the dipole moment can be related to the temperature change ΔT by

$$\Delta P_i = p_i \Delta T \tag{3.8}$$

where the p_i are called the pyroelectric coefficients. Since T is a scalar the p_i form a polar vector. Thus for both ferroelectricity and pyroelectricity the tensor $d_{ijk \ldots n}$ is simply a polar vector and therefore these two phenomena will be possible for any crystal with a point group **G** for which eqn. (3.6) does not lead to the vanishing of all the components of a polar tensor of rank 1, see table 7. Another closely related phenomenon is that of piezo-electricity, in which a crystal may develop an electric dipole moment, **P**, as

Table 7.   The 31 point groups in which ferroelectricity is possible

Type I	Type II	Type III	
1	1′		
2	21′	2′	
m	m1′	m′	
mm2	mm21′	m′m2′	m′m′2
4	41′	4′	
4mm	4mm1′	4′mm′	4m′m′
3	31′		
3m	3m1′	3m′	
6	61′	6′	
6mm	6mm1′	6′mm′	6m′m′

a result of an applied stress. We recall that a stress is described by a tensor of rank 2, with one suffix to describe the direction in which the force is applied and the other suffix to describe the orientation of the plane on which the force is considered to act. The induced electric dipole moment, P_i, is related to the stress, σ_{ij}, by

$$P_i = \sum_{j,k} d_{ijk}\sigma_{jk} \tag{3.9}$$

where the d_{ijk} are called the piezoelectric moduli and form a polar tensor of rank 3. The simplified forms of this tensor for the various point groups are given in table 8 of the book by Nye (1957).

3.2. Equilibrium properties of magnetic crystals

If we consider a crystal of which the symmetry has to be described not by one of the classical point groups but by one of the Heesch–Shubnikov point groups, exactly half of the symmetry operations of the crystal will involve θ, the operation of time-inversion. For each of the symmetry operations R in the unitary subgroup **H** of a Heesch–Shubnikov point group we can use eqns. (3.6) and (3.7). That is, if **M** is a magnetic point group one can immediately make some simplification in the form of a tensor $d_{ijk \ldots n}$ by using the conventional arguments outlined in § 3.1 for the unitary subgroup of **M**. Let us suppose that this has already been done. It then only remains to consider the anti-unitary elements of **M**. Since we can write

$$\mathbf{M} = \mathbf{H} + A\,\mathbf{H} \tag{1.11}$$

where **H** is the unitary subgroup and A is any one of the anti-unitary elements of **M**, we can regard A as the one additional generating element that is needed to generate **M** from **H**. Having found the symmetry properties of $d_{ijk \ldots n}$ in **H**, it is only necessary to consider the additional effect of the element A to find the symmetry properties of $d_{ijk \ldots n}$ in **M**. In this section we describe how this is done for macroscopic static properties of magnetic crystals (Birss 1963, 1964).

The symmetry of a crystal that is not magnetically ordered is, strictly speaking, described by a grey point group $\mathbf{G} + \theta\mathbf{G}$ so that θ itself is a symmetry operation of the crystal and therefore also, as a result of Neumann's principle, of the tensor $d_{ijk \ldots n}$. That is, for a crystal which is not magnetically ordered the tensor $d_{ijk \ldots n}$ is invariant under θ, the operation of time inversion. If a magnetically ordered crystal belongs to one of the 58 type III Heesch–Shubnikov point groups, then θ is no longer a symmetry operation of the crystal on its own but only in combination with certain point-group operations. However, θ^2 is still a symmetry operation of the crystal and this means that θ must have the effect of multiplying $d_{ijk \ldots n}$ by either $+1$ or -1. Therefore, when considering a tensor describing a macroscopic static property of a magnetically ordered crystal, we may distinguish between i-tensors and c-tensors; an i-tensor is one that is invariant under θ, the operation of time-inversion while a c-tensor is one that changes sign under θ. Suppose that θS is any symmetry operation of the set $A\,\mathbf{H}$ of anti-unitary elements of **M**. For an i-tensor we can then replace R by S in either eqn. (3.6) or eqn. (3.7), according as $d_{ijk \ldots n}$ is a polar tensor or an axial tensor, respectively, so that

$$d_{ijk \ldots n} = \sum_{p,\,q,\,r,\,\ldots,\,u} S_{ip} S_{jq} S_{kr} \cdots S_{nu} d_{pqr \ldots u} \tag{3.10}$$

for a polar i-tensor and

$$d_{ijk \ldots n} = |\mathbf{S}| \sum_{p,\,q,\,r,\,\ldots,\,u} S_{ip} S_{jq} S_{kr} \cdots S_{nu} d_{pqr \ldots u} \tag{3.11}$$

for an axial i-tensor where θS is any one of the anti-unitary elements of **M**. However, for a c-tensor we have to replace eqns. (3.10) and (3.11) by

$$d_{ijk \ldots n} = (-1) \sum_{p,\,q,\,r,\,\ldots,\,u} S_{ip} S_{jq} S_{kr} \cdots S_{nu} d_{pqr \ldots u} \tag{3.12}$$

for a polar c-tensor and

$$d_{ijk \ldots n} = (-1)|\mathbf{S}| \sum_{p,\,q,\,r,\,\ldots,\,u} S_{ip} S_{jq} S_{kr} \cdots S_{nu} d_{pqr \ldots u} \tag{3.13}$$

for an axial c-tensor. All the possible simplification of the form of a tensor $d_{ijk \dots n}$ for a type III Heesch–Shubnikov group, **M**, can therefore be achieved by first simplifying $d_{ijk \dots n}$ using the unitary subgroup, **H**, of **M**, and then using the appropriate one of eqns. (3.10)–(3.13) with θS as A, the additional symmetry operation that generates **M** from **H**. A point which sometimes leads to confusion is the identification of the effects of I, θ and θI on a tensor of rank n ; this may be summarized as follows :

	n even				n odd			
	i-tensor		c-tensor		i-tensor		c-tensor	
	polar	axial	polar	axial	polar	axial	polar	axial
I	+	−	+	−	−	+	−	+
θ	+	+	−	−	+	+	−	−
θI	+	−	−	+	−	+	+	−

A comprehensive account of the simplification of the form of $d_{ijk \dots n}$ for tensors corresponding to various physical properties of magnetically ordered crystals has been given by Birss (1963, 1964). For example, the groups in which a non-zero polar vector can exist, and therefore in which ferroelectricity and pyroelectricity are possible, are identified in table 7. The point groups in which a non-zero axial vector can exist, and therefore in which ferromagnetism is possible, are identified in table 8. One other example which is of considerable importance is the magnetoelectric tensor. The simplest magnetoelectric effect is the appearance, in a crystal which is not spontaneously magnetized, of a magnetic moment M_i when the crystal is situated in an electric field E_i. M_i and E_i are related by

$$M_i = \sum_j Q_{ij} E_j \qquad (3.14)$$

Table 8. The 31 point groups in which ferromagnetism is possible

Type I			Type III						
1	$\bar{1}$								
2	m	$2/m$	$2'$	m'	$2'/m'$	$22'2'$	$m'm'2$	$m'm2'$	$mm'm'$
3	$\bar{3}$		$32'$	$3m'$	$\bar{3}m'$				
4	$\bar{4}$	$4/m$	$42'2'$	$4m'm'$	$\bar{4}2'm'$	$4/mm'm'$			
6	$\bar{6}$	$6/m$	$62'2'$	$6m'm'$	$\bar{6}m'2'$	$6/mm'm'$			

Table 9. The magnetoelectric tensor

Magnetic point group	Q_{ij}
$1,\ \bar{1}'$	$\begin{bmatrix} Q_{11} & Q_{12} & Q_{13} \\ Q_{21} & Q_{22} & Q_{23} \\ Q_{31} & Q_{32} & Q_{33} \end{bmatrix}$
$2,\ m',\ 2/m'$	$\begin{bmatrix} Q_{11} & Q_{12} & 0 \\ Q_{21} & Q_{22} & 0 \\ 0 & 0 & Q_{33} \end{bmatrix}$
$2',\ m,\ 2'/m$	$\begin{bmatrix} 0 & 0 & Q_{13} \\ 0 & 0 & Q_{23} \\ Q_{31} & Q_{32} & 0 \end{bmatrix}$
$222,\ m'm'2,\ m'm'm'$	$\begin{bmatrix} Q_{11} & 0 & 0 \\ 0 & Q_{22} & 0 \\ 0 & 0 & Q_{33} \end{bmatrix}$
$22'2',\ mm2,\ (m'm2'),\ m'mm$	$\begin{bmatrix} 0 & Q_{12} & 0 \\ Q_{21} & 0 & 0 \\ 0 & 0 & 0 \end{bmatrix}$
$4,\ \bar{4}',\ 4/m',\ 3,\ \bar{3}',\ 6,\ \bar{6}',\ 6/m'$	$\begin{bmatrix} Q_{11} & Q_{12} & 0 \\ -Q_{12} & Q_{11} & 0 \\ 0 & 0 & Q_{33} \end{bmatrix}$
$4',\ \bar{4},\ 4'/m'$	$\begin{bmatrix} Q_{11} & Q_{12} & 0 \\ Q_{12} & -Q_{11} & 0 \\ 0 & 0 & 0 \end{bmatrix}$
$422,\ 4m'm',\ \bar{4}'2m',\ 4/m'm'm',\ 32,\ 3m',$ $\bar{3}'m',\ 622,\ 6m'm',\ \bar{6}'m'2,\ 6/m'm'm'$	$\begin{bmatrix} Q_{11} & 0 & 0 \\ 0 & Q_{11} & 0 \\ 0 & 0 & Q_{33} \end{bmatrix}$
$4'22',\ (4'mm'),\ \bar{4}2m,\ (\bar{4}2'm'),\ 4'/m'mm'$	$\begin{bmatrix} Q_{11} & 0 & 0 \\ 0 & -Q_{11} & 0 \\ 0 & 0 & 0 \end{bmatrix}$
$42'2',\ 4mm,\ \bar{4}'2'm,\ 4/m'mm,\ 32',\ 3m,$ $\bar{3}'m,\ 62'2',\ 6mm,\ \bar{6}'m2',\ 6/m'mm$	$\begin{bmatrix} 0 & Q_{12} & 0 \\ -Q_{12} & 0 & 0 \\ 0 & 0 & 0 \end{bmatrix}$
$23,\ m'3,\ 432,\ \bar{4}'3m',\ m'3m'$	$\begin{bmatrix} Q_{11} & 0 & 0 \\ 0 & Q_{11} & 0 \\ 0 & 0 & Q_{11} \end{bmatrix}$

(Indenbom 1960)

where Q_{ij} is a tensor of rank 2 called the magnetoelectric tensor. Since M_i is an axial c-tensor of rank 1 and E_i is a polar i-tensor of rank 1, the magneto-electric tensor Q_{ij} must be an axial c-tensor of rank 2. The form of Q_{ij} is given in table 9 for those point groups for which Q_{ij} is not null (Indenbom 1960, Birss 1964). The existence of magnetoelectric effects was first pre-dicted by Curie towards the end of the nineteenth century but no conclusive observation of a magnetoelectric effect was obtained until the work of Astrov (1960) using antiferromagnetic Cr_2O_3. Some of the earlier attempts to observe magnetoelectric effects had been unsuccessful because people had used materials for which the tensor Q_{ij} is null. Apart from its intrinsic interest the magnetoelectric effect has been useful in certain cases as an aid to the determination of magnetic structures, see § 8.1. An extensive discussion of magnetoelectric effects will be found in the book by O'Dell (1967).

3.3. *Transport properties*

In the discussion that we have given in the two previous sections it has been assumed that the tensor $d_{ijk \ldots n}$ is being used to describe a static or equilibrium property of a crystalline material. There are certain complications which arise when we come to consider tensors representing transport properties such as the electrical conductivity or the thermal conductivity because these transport coefficients involve the time, t, implicitly at least. The situation is slightly different for non-magnetic crystals and for magnetically ordered crystals and so we shall consider these two cases in turn.

For a crystal that is not magnetically ordered the discussion that we gave in § 3.1 based on eqns. (3.6) and (3.7), using crystallographic symmetry and Neumann's principle, can also be used for tensors representing transport properties. It is when we come to consider the intrinsic symmetry, as distinct from the crystallographic symmetry, of the tensors representing transport properties that we find differences from the case of the static, or equilibrium, properties. The discussion of the intrinsic symmetry of tensors representing transport properties is based on the use of a theorem due to Onsager (1931 a, b) which is a result of thermodynamical arguments. We do not propose to consider all the details here (see, for example, De Groot (1951), Nye (1957)). We simply give the results. If σ_{ij} and κ_{ij} are the tensors representing the electrical conductivity and the thermal conductivity, respectively, then Onsager's work showed that these tensors must be symmetric tensors, namely

$$\left.\begin{array}{c} \sigma_{ij} = \sigma_{ji} \\ \\ \kappa_{ij} = \kappa_{ji} \end{array}\right\} . \tag{3.15}$$

Moreover, Onsager's work also led to the establishment of a relationship between the Peltier tensor, π_{ij}, and the Seebeck tensor, α_{ij}, namely

$$\frac{1}{T} \pi_{ij} = \alpha_{ji}. \tag{3.16}$$

The conditions expressed in eqns. (3.15) and (3.16) represent intrinsic symmetry properties of these tensors and apply regardless of any considerations of crystallographic symmetry.

If we wish to determine the restrictions that are imposed by symmetry on the form of a tensor that represents one of the transport properties of a non-magnetic crystal the procedure is relatively straightforward. There are two separate stages :

  (i) Use the results of Onsager's theorem as given in eqns. (3.15) and (3.16), and

(ii) Use eqn. (3.6) or eqn. (3.7), as appropriate, based on crystallographic symmetry and the use of Neumann's principle, with all the generating elements of the point group **G** to simplify the form of the tensor.

These two rules can be applied in either order.

In the case of a crystal which is magnetically ordered, or if a non-magnetic crystal is placed in an external magnetic field, so that its symmetry is described by one of the non-unitary Heesch–Shubnikov point groups, the argument given above needs to be modified. With regard to (i), the form of the results of the use of Onsager's theorem will be slightly different. If a crystal is subjected to a magnetic field, **H**, that may be an external field or may arise as an internal field in the crystal, it is possible to show (Onsager 1931 a, b, De Groot 1951, Nye 1957, Birss 1964, Kleiner 1966) that eqns. (3.15) and (3.16) have to be replaced by

$$\left.\begin{aligned}\sigma_{ij}(\mathbf{H}) &= \sigma_{ji}(-\mathbf{H}) \\ \kappa_{ij}(\mathbf{H}) &= \kappa_{ji}(-\mathbf{H})\end{aligned}\right\} \tag{3.17}$$

and

$$\frac{1}{T}\,\pi_{ij}(\mathbf{H}) = \alpha_{ji}(-\mathbf{H}) \tag{3.18}$$

respectively. Thus rule (i) becomes modified by the use of eqns. (3.17) and (3.18) instead of eqns. (3.15) and (3.16). Rule (ii) can still be used, as before, but now only for the elements of the unitary subgroup **H** of **M** (Birss 1964) ; the rule has to be modified when it comes to using the anti-unitary elements in the set $A\,\mathbf{H}$ and we shall see how this comes about. Suppose that the anti-unitary elements in $A\,\mathbf{H}$ are of the form θS, where S is a point-group operation. Let us suppose that Neumann's principle also holds for these operations θS. We should then have for the electrical conductivity, for example

$$(\theta S)\sigma_{ij}(\mathbf{H}) = \sigma_{ij}(\mathbf{H}). \tag{3.19}$$

After some manipulation it is possible to show that this leads to

$$\sigma_{ij}(\mathbf{H}) = \sum_{p,\,q} S_{ip}S_{jq}\sigma_{pq}{}^*(\mathbf{H}) \tag{3.20}$$

(for details see Cracknell (1973 d)). This equation can be generalized to a tensor of arbitrary rank so that

$$d_{ijk\,\ldots\,n}(\mathbf{H}) = \sum_{p,\,q,\,r,\,\ldots,\,u} S_{ip}S_{jq}S_{kr}\ldots S_{nu}d^*{}_{pqr\,\ldots\,u}(\mathbf{H}) \tag{3.21}$$

for a polar tensor and

$$d_{ijk\,\ldots\,n}(\mathbf{H}) = |\mathbf{S}| \sum_{p,\,q,\,r,\,\ldots,\,u} S_{ip}S_{jq}S_{kr}\ldots S_{nu}d^*{}_{pqr\,\ldots\,u}(\mathbf{H}) \tag{3.22}$$

for an axial tensor. Thus for an anti-unitary element θS in a type III Heesch–Shubnikov point group we use eqn. (3.21) or (3.22) to simplify the form of a property tensor in a similar manner to the use of eqn. (3.6) or (3.7) for the unitary elements. Therefore, to summarize, the prescription for simplifying the form of a tensor that represents a transport coefficient for a magnetic crystal is :

(i) Use eqns. (3.17) and (3.18) based on Onsager's theorem,
(ii) Use eqn. (3.6) or (3.7) based on Neumann's principle for the unitary elements, R, of **M**,
(iii) Use eqn. (3.21) or (3.22) based on a modified form of Neumann's principle, for the anti-unitary elements θS of **M**.

The result of the application of these rules to tensors for the electrical conductivity, the thermal conductivity, and the thermoelectric effects is shown in table 10. To illustrate the use of table 10 we consider the extra-ordinary Hall effect in a ferromagnetic metal, that is the contribution to the Hall effect arising from the internal field in such a metal. Let us consider the example of ferromagnetic Co, which has the h.c.p. structure. For the existence of the extraordinary Hall effect, at least some of the off-diagonal components of $\sigma_{ij}(\mathbf{H})$ must be non-zero. For the various possible orienta-tions of **M**, the magnetization, relative to the crystallographic axes, the appropriate magnetic point groups will be

$$\mathbf{M}\|[0001] \qquad 6/mm'm'$$

$$\mathbf{M}\|[10\bar{1}0] \qquad mm'm'$$

$$\mathbf{M}\|[11\bar{2}0] \qquad m'mm'$$

$$\mathbf{M}\|[uv\dagger0] \qquad 2'/m'$$

$$\mathbf{M}\|[uv\dagger w] \qquad \bar{1}.$$

From table 10 we see that, for an arbitrary orientation of the magnetization, relative to the crystallographic axes, the extraordinary Hall effect is permitted by symmetry considerations ; indeed it also appears to have been observed in practice (Cheremushkina and Vasil'eva 1966, Yu and Chang 1970). How-ever, the off-diagonal elements vanish for a single-domain specimen of ferro-magnetic Co that is magnetized parallel to [0001], [10$\bar{1}$0] or [11$\bar{2}$0], see table 10, and so on symmetry grounds the extraordinary Hall effect is not allowed for these special magnetization directions (for further details see Cracknell (1973 d)).

Our discussion of transport coefficients has been based on a macroscopic approach. It is also possible to consider this problem from a microscopic viewpoint (see Kleiner (1966, 1967, 1969), Cracknell (1973 d)). We shall not consider the details of the microscopic approach here, except to note one point. This concerns the fact that the treatment given by Kleiner involves the use of a group which contains more symmetry than the crystal possesses. In addition to all the symmetry operations that leave invariant $\mathscr{H}(\mathbf{H})$, the Hamiltonian of a magnetic crystal or of a non-magnetic crystal situated in an external magnetic field, Kleiner also included those operations that send $\mathscr{H}(\mathbf{H})$ into $\mathscr{H}(-\mathbf{H})$. In this formalism the Onsager relations are not used explicitly ; instead, they arise from the use of the transformation equations for a tensor under the operations of this larger symmetry group.

Table 10. Forms of tensors for thermogalvanomagnetic coefficients for type III Heesch–Shubnikov point groups

Point groups	$\sigma_{ij}(\mathbf{H})$	$\alpha_{ij}(\mathbf{H})$
Triclinic : $\bar{1}'$	$\begin{pmatrix} \sigma_{11}(\mathbf{H}^2) & \sigma_{12}(\mathbf{H}) & \sigma_{13}(\mathbf{H}) \\ \sigma_{12}(-\mathbf{H}) & \sigma_{22}(\mathbf{H}^2) & \sigma_{23}(\mathbf{H}) \\ \sigma_{13}(-\mathbf{H}) & \sigma_{23}(-\mathbf{H}) & \sigma_{33}(\mathbf{H}^2) \end{pmatrix}$	$\begin{pmatrix} \alpha_{11}(\mathbf{H}) & \alpha_{12}(\mathbf{H}) & \alpha_{13}(\mathbf{H}) \\ \alpha_{21}(\mathbf{H}) & \alpha_{22}(\mathbf{H}) & \alpha_{23}(\mathbf{H}) \\ \alpha_{31}(\mathbf{H}) & \alpha_{32}(\mathbf{H}) & \alpha_{33}(\mathbf{H}) \end{pmatrix}$
Monoclinic : $2'$; m' ; $2/m'$; $2'/m$; $2'/m'$	$\begin{pmatrix} \sigma_{11}(\mathbf{H}^2) & \sigma_{12}(\mathbf{H}) & 0 \\ \sigma_{12}(-\mathbf{H}) & \sigma_{22}(\mathbf{H}^2) & 0 \\ 0 & 0 & \sigma_{33}(\mathbf{H}^2) \end{pmatrix}$	$\begin{pmatrix} \alpha_{11}(\mathbf{H}) & \alpha_{12}(\mathbf{H}) & 0 \\ \alpha_{21}(\mathbf{H}) & \alpha_{22}(\mathbf{H}) & 0 \\ 0 & 0 & \alpha_{33}(\mathbf{H}) \end{pmatrix}$
Orthorhombic : $2'2'2$; $m'm'2$; $m'm2'$; mmm' ; $m'm'm$	$\begin{pmatrix} \sigma_{11}(\mathbf{H}^2) & 0 & 0 \\ 0 & \sigma_{22}(\mathbf{H}^2) & 0 \\ 0 & 0 & \sigma_{33}(\mathbf{H}^2) \end{pmatrix}$	$\begin{pmatrix} \alpha_{11}(\mathbf{H}) & 0 & 0 \\ 0 & \alpha_{22}(\mathbf{H}) & 0 \\ 0 & 0 & \alpha_{33}(\mathbf{H}) \end{pmatrix}$
Trigonal, tetragonal, and hexagonal[a]	$\begin{pmatrix} \sigma_{11}(\mathbf{H}^2) & 0 & 0 \\ 0 & \sigma_{11}(\mathbf{H}^2) & 0 \\ 0 & 0 & \sigma_{33}(\mathbf{H}^2) \end{pmatrix}$	$\begin{pmatrix} \alpha_{11}(\mathbf{H}) & 0 & 0 \\ 0 & \alpha_{11}(\mathbf{H}) & 0 \\ 0 & 0 & \alpha_{33}(\mathbf{H}) \end{pmatrix}$
Cubic : $m'3$; $\bar{4}'3m'$; $4'32'$; $m'3m$; $m'3m'$	$\begin{pmatrix} \sigma_{11}(\mathbf{H}^2) & 0 & 0 \\ 0 & \sigma_{11}(\mathbf{H}^2) & 0 \\ 0 & 0 & \sigma_{11}(\mathbf{H}^2) \end{pmatrix}$	$\begin{pmatrix} \alpha_{11}(\mathbf{H}) & 0 & 0 \\ 0 & \alpha_{11}(\mathbf{H}) & 0 \\ 0 & 0 & \alpha_{11}(\mathbf{H}) \end{pmatrix}$

[a] $4'$; $\bar{4}'$; $42'2'$; $4/m'$; $4'/m$; $4m'm'$; $4'mm'$; $\bar{4}2m'$; $\bar{4}'m2'$; $4/m'm'm'$; $4/m'mm'$; $4'/m'm'm$; $4'/mmm$; $4/m'mm$; $4/mm'm$; $6'2'2$; $6/m'$; $6'/m'$; $\bar{6}'$; $\bar{6}m'2$; $6'$; $\bar{3}'$; $3m'$; $3'm$; $62'2'$; $6/m'$; $6'/m'$; $6m'm'$; $6'mm'$; $\bar{6}'m'm$; $6/m'm'm$; $6/mm'm'$.

Notes. (i) The elements of each of the type III Heesch–Shubnikov point groups can be identified, for example, from table 7.1 of Bradley and Cracknell (1972). (ii) We have restricted the tensor components to being real. (iii) Only the forms of $\sigma_{ij}(\mathbf{H})$ and $\alpha_{ij}(\mathbf{H})$ are given. The form of $\kappa_{ij}(\mathbf{H})$ is the same as that of $\sigma_{ij}(\mathbf{H})$, while the form of $\pi_{ij}(\mathbf{H})$ can be obtained from $\alpha_{ij}(\mathbf{H})$ via eqn. (3.18). (Cracknell 1973 d).

§ 4. Ligand-field theory

The term 'crystal-field theory', which we used in § 2.1, and the term 'ligand-field theory', which we now use, are not quite synonymous with each other. Nevertheless, the two expressions are often taken to be synonymous in practice, although ligand-field theory is commonly regarded as being more general than crystal-field theory. As initiated by Bethe (1929) crystal-field theory was concerned with the splitting of an atomic or ionic energy level of given l value by an electrostatic field with the symmetry of one of the crystallographic point groups. Consider an ionic crystal that is composed of two constituents X^+ and Y^- and suppose that the energy levels and wave functions of a free X^+ or Y^- ion are known in detail. In crystal-field theory we study the energy levels and wave functions of an ion in the crystal of XY and, in particular, we try to relate these wave functions and energy levels to those for a free ion of the same element. The ion under consideration may be one of the X^+ or Y^- ions of which the crystal is constituted. Alternatively, it may be some impurity ion which may be in a substitutional or interstitial position in the structure. In crystal-field theory we impose a rigid separation between the ion under investigation and the whole of the rest of the crystal. That is, we think of the rest of the ions in the crystal as an array of point charges whose function is merely to produce an electrostatic field, with a certain point-group symmetry, at the site occupied by our specimen ion. Similarly, we suppose that the only effect of the crystalline environment on the specimen ion is that it experiences a perturbation which is the electrostatic field produced by all the other ions in the crystal. In this electrostatic field the energy levels of the free ion can be expected to split in the manner we have outlined in § 2.1. This approach is unrealistically restrictive. The more general approach of ligand-field theory allows some electron transfer to occur between a specimen ion and the surrounding ions in the crystal. In terms of bonding mechanisms this means that ligand-field theory allows one to take into account the well-known fact that in any supposedly ionic crystal there is some degree of covalent bonding. We shall not pursue this largely semantic discussion of the difference between crystal-field theory and ligand-field theory any further.

In § 2.1 we were only concerned with a qualitative discussion of the splittings of atomic energy levels and we were not concerned with the magnitudes of the splittings. In this part we shall consider the extension of the argument to the inclusion of spin and to some discussion of the use of symmetry arguments to simplify the calculations of the magnitudes of the splittings. This involves the use of symmetry arguments to determine the forms of the wave functions of atoms or ions in crystals. In § 2.1 we were not very careful to distinguish between one-electron wave functions and many-electron wave functions and talked indiscriminately about l for a single electron or L for a many-electron configuration. That is, we assumed that L remained a good quantum number for an ion in a crystal. In § 4.1 we shall restrict ourselves to one-electron wave functions, that is we assume that we have an ion in which the electronic structure consists entirely of closed shells apart from a single outermost electron. The effects of the inclusion of electron spin and of magnetic ordering, still for one-electron wave functions, will be discussed in §§ 4.2 and 4.5, respectively. We shall consider many-electron wave functions in § 4.3.

4.1. *One-electron wave functions*

In calculations of atomic structures it is conventional to express the wave functions of the individual electrons in terms of spherical polar coordinates r, θ and ϕ. If one assumes that it is reasonable to use a one-electron Schrödinger equation for any given electron in an atom and that the potential is spherically symmetrical, the wave function can be written as the product of a radial function $R_{nl}(r)$ and a spherical harmonic $Y_l^m(\theta, \phi)$, that is

$$\psi_{nl}^m(r, \theta, \phi) = R_{nl}(r) Y_l^m(\theta, \phi). \tag{4.1}$$

Although the existence of spherical symmetry is not strictly true, except perhaps for atoms of hydrogen, nevertheless it is well known that a meaningful account of the structures of atoms and of the periodic table of the chemical elements can be obtained in terms of the quantum numbers n, l and m_l, where we have $m_l = m$. This indicates that functions of the form in eqn. (4.1) give very good approximations to the true wave functions of electrons in a free atom or ion. For any atom more complicated than hydrogen $R_{nl}(r)$ is a function that has to be determined by a numerical integration of the appropriate radial equation. However, the spherical harmonics are well-known analytical functions that can be evaluated and tabulated for any required sets of θ and ϕ.

For an electron which is localized on an ion in a crystal, the potential is no longer spherically symmetrical and so it is no longer possible to express $\psi(r, \theta, \phi)$ as a single term in the form of eqn. (4.1). However, it is still reasonable to expand $\psi(r, \theta, \phi)$ as a series of such terms, that is as a linear combination of atomic orbitals (LCAO), so that we can write

$$\psi(r, \theta, \phi) = \sum_{n, l, m} A_{nl}^m R_{nl}(r) Y_l^m(\theta, \phi). \tag{4.2}$$

There are a large number of coefficients A_{nl}^m that would need to be determined in a completely general expansion for $\psi(r, \theta, \phi)$. But in practice it will be possible to eliminate all, or nearly all, of them for various reasons. On physical grounds it is reasonable to expect that the most important terms in the expansion of $\psi(r, \theta, \phi)$ in eqn. (4.2) for a given electron will be those arising from eigenstates that are close in energy to the related eigenstate of the given electron in the free atom or ion. Thus one is likely to need to include only one value of n, and probably also only one value of l, in the expansion of $\psi(r, \theta, \phi)$ in eqn. (4.2). In a rough and ready manner we can say that this restriction corresponds to the crystal-field approximation in which electron transfer is not allowed ; if the possibility of electron transfer is allowed (ligand-field theory) functions with other l and n values may be included in the expansion of $\psi(r, \theta, \phi)$. We have assumed that the eigenstates of the atom or ion in the crystal can be regarded, formally at least, as emerging from the eigenstates of the free atom or ion by slowly ' turning on ' the potential or field due to the rest of the crystal.

In addition to restricting the summation in eqn. (4.2) on physical grounds it may be possible to achieve a further simplification by making use of the symmetry of the local environment of an atom or ion in a crystal. We have not included the spin parts of the wave functions of the electrons so that we have the situation described as (A) in § 2.1, namely a crystal field with the symmetry of a classical point group **G** in the absence of spin–orbit coupling. That is, the potential possesses the symmetry of the point group **G**. We have seen that the eigenstates of the localized electrons in an atom or ion in a crystal can be classified according to the single-valued irreducible representations Γ_i of **G**. For an electron in a state labelled by Γ_i the wave function $\psi(r, \theta, \phi)$ must transform according to Γ_i under the operations of the point group **G** ; that is $\psi(r, \theta, \phi)$ must be a basis of Γ_i. By imposing this condition on $\psi(r, \theta, \phi)$ it may be possible to show that some of the coefficients $A_{nl}{}^m$ must vanish or that some of the coefficients are linearly related to one another. Such functions can be generated by forming the function

$$W^i Y_l{}^m(\theta, \phi) = \sum_R \chi^i(R)^* R Y_l{}^m(\theta, \phi) \tag{4.3}$$

where the summation over R is over the elements of the group **G** (see eqn. (1.10)). For each of the 32 ordinary point groups, **G**, which are unitary groups, the functions that are symmetry-adapted to all the various irreducible representations of **G** have been determined by this method (see, for example, Altmann (1957), Altmann and Bradley (1963), Altmann and Cracknell (1965), Bradley and Cracknell (1972)). For non-cubic point groups $W^i Y_l{}^m(\theta, \phi)$ can be expressed in a simple general form for any values of l and m ; an example is given in table 11. However, for the cubic point groups $W^i Y_l{}^m(\theta, \phi)$ will generally be a linear combination of several spherical harmonics, all with

Table 11.   Symmetry-adapted functions for the point group 222 (D_2)

	l	m	ϕ-dep
Γ_1	0	0	c
	3	2	s
Γ_2	2	1	c
	1	1	s
Γ_3	2	2	s
	1	0	c
Γ_4	2	1	s
	1	1	c

Notes.   (i)

$$Y_l{}^{m,\, c}(\theta, \phi) = \{Y_l{}^m(\theta, \phi) + Y_l{}^{-m}(\theta, \phi)\}/\sqrt{2},$$
$$Y_l{}^{m,\, s}(\theta, \phi) = -i\{Y_l{}^m(\theta, \phi) - Y_l{}^{-m}(\theta, \phi)\}/\sqrt{2},$$

which are real and have ϕ-dependence $\cos(m\phi)$ and $\sin(m\phi)$ respectively. (ii) Multiples of 2 can be added to, but not subtracted from, the values of l and m given.

Table 12. Symmetry-adapted functions for the one-dimensional representations of the cubic point groups

23 (T)	m3 (T_h)	43m (T_d)	432 (O)	m3m (O_h)	l	φ-dep	Function
A	A_g	A_1	A_1	A_{1g}	0	—	1(0)
A	A_u	A_1	A_2	A_{2u}	3	s	1(2)
A	A_g	A_1	A_1	A_{1g}	4	c	0.76376261583(0)+0.64549722437(4)
A	A_g	A_1	A_1	A_{1g}	6	c	0.35355339059(0)−0.93541434669(4)
A	A_g	A_2	A_2	A_{2g}	6	c	0.82915619759(2)−0.55901699438(6)
A	A_u	A_2	A_2	A_{2u}	7	s	0.73598007219(2)+0.67700320038(6)
A	A_g	A_1	A_1	A_{1g}	8	c	0.71807033082(0)+0.38188130791(4)+0.58184333516(8)
A	A_u	A_1	A_2	A_{2u}	9	s	0.43301270189(2)−0.90138781887(6)
A	A_u	A_2	A_1	A_{1u}	9	s	0.84162541153(4)−0.54006172487(8)
A	A_g	A_1	A_1	A_{1g}	10	c	0.41142536788(0)−0.58630196998(4)−0.69783892602(8)
A	A_g	A_2	A_2	A_{2g}	10	c	0.80201568979(2)+0.15728821740(6)−0.57622152858(10)
A	A_u	A_1	A_2	A_{2u}	11	s	0.66536330928(2)+0.45927932677(6)+0.58851862049(10)
A	A_g	A_1	A_1	A_{1g}	12	c	0.69550266594(0)+0.31412556680(4)+0.34844953759(8)+0.54422797585(12)
A	A_g	A_1	A_1	A_{1g}	12	c	0.55897937420(4)−0.80626750818(8)+0.19358399848(12)
A	A_g	A_2	A_2	A_{2g}	12	c	0.21040635288(2)−0.82679728471(6)+0.52166600107(10)

Note. The functions are given as linear combinations of $Y_l^{m,c}(\theta, \phi)$ or of $Y_l^{m,s}(\theta, \phi)$, where the appropriate value of m is given in brackets following each coefficient in the table. For further details see chapter 2 of Bradley and Cracknell (1972). For 432 (O) A_1 and A_2 are alternative labels for Γ_1 and Γ_2, respectively, which were used in table 2.

the original value of l but with various different values of m. The coefficients in this expression for a cubic point group are complicated and the expansion cannot be expressed in a simple general form; an example is given in table 12.

Table 13. Compatibilities between the irreducible representations of the point groups $m3m$ (O_h), 432 (O) and 222 (D_2)

$m3m$	432	222
$\Gamma_1^\pm$	Γ_1	Γ_1
$\Gamma_2^\pm$	Γ_2	Γ_1
$\Gamma_3^\pm$	Γ_3	$2\Gamma_1$
$\Gamma_4^\pm$	Γ_4	$\Gamma_2 \oplus \Gamma_3 \oplus \Gamma_4$
$\Gamma_5^\pm$	Γ_5	$\Gamma_2 \oplus \Gamma_3 \oplus \Gamma_4$
$\Gamma_6^\pm$	Γ_6	Γ_5
$\Gamma_7^\pm$	Γ_7	Γ_5
$\Gamma_8^\pm$	Γ_8	$2\Gamma_5$

As an example of the use of the tables for non-cubic groups let us consider the splitting of an atomic or ionic F term in a crystal field with the symmetry of the point group 222 (D_2). If the atomic F term arises from a single electron and all the other electrons in the atom are in closed shells, the wave function of this electron ($l=3$) may be any one of the seven ($=(2l+1)$) degenerate wave functions

$$\psi_{n3}{}^m(r, \theta, \phi) = R_{n3}(r) Y_3{}^m(\theta, \phi) \qquad (4.4)$$

where n is the principal quantum number, $m = -3, -2, -1, 0, 1, 2$ or 3, and the radial function $R_{n3}(r)$ is, in principle, known. By using the compatibility tables between the irreducible representations of the rotation group $O(3)$ and of 222, see table 13, we find that in an environment with the symmetry of this point group the F term is split into seven non-degenerate levels according to

$$\mathscr{D}^3 = \Gamma_1 \oplus 2\Gamma_2 \oplus 2\Gamma_3 \oplus 2\Gamma_4. \qquad (4.5)$$

We shall assume that the wave functions of these terms involve mixing only the seven wave functions in eqn. (4.4). If we also assume that the radial part of each wave function is the same in the solid as in the free atom or ion, the wave functions for each of these terms in the solid can be completely determined. From table 11 we can see how the spherical harmonics with $l=3$ are distributed among the irreducible representations of the point group 222 :

$$
\left.
\begin{array}{ll}
\Gamma_1 & Y_3{}^{2,\,s}(\theta, \phi) \\[2mm]
\Gamma_2 & Y_3{}^{1,\,s}(\theta, \phi),\ Y_3{}^{3,\,s}(\theta, \phi) \\[2mm]
\Gamma_3 & Y_3{}^{0,\,c}(\theta, \phi),\ Y_3{}^{2,\,c}(\theta, \phi) \\[2mm]
\Gamma_4 & Y_3{}^{1,\,c}(\theta, \phi),\ Y_3{}^{3,\,c}(\theta, \phi)
\end{array}
\right\}. \qquad (4.6)
$$

Using the definitions of $Y_l^{m,\,c}(\theta,\,\phi)$ and $Y_l^{m,\,s}(\theta,\,\phi)$ this means that the seven wave functions will be

$$
\left.
\begin{array}{ll}
\Gamma_1 & 2^{-1/2}R_{n3}(r)\{Y_3^{\,2}(\theta,\,\phi) - Y_3^{\,-2}(\theta,\,\phi)\} \\[2mm]
\Gamma_2 & 2^{-1/2}R_{n3}(r)\{Y_3^{\,1}(\theta,\,\phi) - Y_3^{\,-1}(\theta,\,\phi)\} \\[2mm]
& 2^{-1/2}R_{n3}(r)\{Y_3^{\,3}(\theta,\,\phi) - Y_3^{\,-3}(\theta,\,\phi)\} \\[2mm]
\Gamma_3 & R_{n3}(r)\,Y_3^{\,0}(\theta,\,\phi) \\[2mm]
& 2^{-1/2}R_{n3}(r)\{Y_3^{\,2}(\theta,\,\phi) + Y_3^{\,-2}(\theta,\,\phi)\} \\[2mm]
\Gamma_4 & 2^{-1/2}R_{n3}(r)\{Y_3^{\,1}(\theta,\,\phi) + Y_3^{\,-1}(\theta,\,\phi)\} \\[2mm]
& 2^{-1/2}R_{n3}(r)\{Y_3^{\,3}(\theta,\,\phi) + Y_3^{\,-3}(\theta,\,\phi)\}
\end{array}
\right\} \quad (4.7)
$$

where we have neglected phase factors of $-i$ in some of these expressions.

As an example of the forms of the wave functions for cubic point groups we consider an F term in a crystal-field environment with the symmetry of the point group 432 (O). Again we assume that this term arises from a single electron and that all the other electrons in the atom are in closed shells. We have already seen in § 2.1 that in a field with the symmetry of the cubic point group 432 an atomic F term splits according to

$$
\mathscr{D}^3 = \Gamma_2 \oplus \Gamma_4 \oplus \Gamma_5 \qquad (4.8)
$$

that is, into one non-degenerate term and two three-fold degenerate terms (Γ_4 and Γ_5), see fig. 7. If we assume, as in the previous example, that the Γ_2, Γ_4 and Γ_5 terms involve the mixing of states from the original F term and no others, the wave functions will be linear combinations of the form given in eqn. (4.2) where the summation is restricted to one value of n, to one value of l ($l=3$) and to $m=-3,\,-2,\,-1,\,0,\,1,\,2$ and 3. From table 2.6 of Bradley and Cracknell (1972), of which an extract has just been reproduced in table 12, one finds that these wave functions must be

$$
\psi_{\Gamma_2}(r,\,\theta,\,\phi) = R_{n3}(r)\,Y_3^{\,2,\,s}(\theta,\,\phi) \qquad (4.9\,a)
$$

or

or

$$
\left.
\begin{array}{l}
\psi_{\Gamma_4}(r,\,\theta,\,\phi) = R_{n3}(r)\{a\,Y_3^{\,1,\,c}(\theta,\,\phi) - b\,Y_3^{\,3,\,c}(\theta,\,\phi)\} \\[2mm]
R_{n3}(r)\{a\,Y_3^{\,1,\,s}(\theta,\,\phi) + b\,Y_3^{\,3,\,s}(\theta,\,\phi)\} \\[2mm]
-\,R_{n3}(r)\,Y_3^{\,0}(\theta,\,\phi)
\end{array}
\right\} \quad (4.9\,b)
$$

or

or

$$
\left.
\begin{array}{l}
\psi_{\Gamma_5}(r,\,\theta,\,\phi) = R_{n3}(r)\{-b\,Y_3^{\,1,\,c}(\theta,\,\phi) - a\,Y_3^{\,3,\,c}(\theta,\,\phi)\} \\[2mm]
R_{n3}(r)\{b\,Y_3^{\,1,\,s}(\theta,\,\phi) - a\,Y_3^{\,3,\,s}(\theta,\,\phi)\} \\[2mm]
R_{n3}(r)\,Y_3^{\,2,\,c}(\theta,\,\phi)
\end{array}
\right\} \quad (4.9\,c)
$$

where $a = 0 \cdot 612\,372\,435\,70$ and $b = 0 \cdot 790\,569\,415\,04$ (to eleven significant figures).

We have illustrated how to construct wave functions for localized electrons in a solid in terms of atomic wave functions. In order to calculate the magnitudes of the crystal-field splittings which occur for a given ion, it is also necessary to know the electrostatic potential experienced by that ion. The form of this potential can also be simplified by using symmetry arguments, as we shall see in § 4.4.

4.2. *The inclusion of spin*

In the previous section we ignored the existence of electron spin. It will be recalled that for a complete description of the behaviour of the electrons in an atom it is necessary to introduce a fourth quantum number m_s in addition to the three quantum numbers n, l and m_l. m_s is the spin quantum number and it takes one of the two values $+\frac{1}{2}$ and $-\frac{1}{2}$. In this section we shall be concerned with the modifications to the theory of one-electron wave functions which become necessary when electron spin is included. If we use $\phi(m_s)$ to denote the spinors

$$\left.\begin{array}{l} \alpha = \begin{pmatrix} 1 \\ 0 \end{pmatrix} \quad \text{if } m_s = -\tfrac{1}{2} \\[20pt] \beta = \begin{pmatrix} 0 \\ 1 \end{pmatrix} \quad \text{if } m_s = +\tfrac{1}{2} \end{array}\right\} \tag{4.10}$$

the wave function of an electron in the state specified by the four quantum numbers n, l, m_l and m_s can be written as

$$\psi_{nl}{}^{mm_s}(r, \theta, \phi) = R_{nl}(r) Y_l{}^m(\theta, \phi)\phi(m_s) \tag{4.11}$$

where we have replaced m_l by m. This equation replaces eqn. (4.1). The one-electron wave functions for such an atom or ion when it is subjected to an environment with the symmetry of a point group **G** can be expanded in terms of wave functions for the free atom or ion

$$\psi(r, \theta, \phi) = \sum_{n, l, m, m_s} A_{nl}{}^{mm_s} R_{nl}(r) Y_l{}^m(\theta, \phi)\phi(m_s) \tag{4.12}$$

which is analogous to that given in eqn. (4.2). As before, it is likely that physical arguments can be used to restrict very considerably the summations over n and l on the right-hand side of eqn. (4.12). The expansion in eqn. (4.12) can be simplified still further by imposing the requirement that it must belong to one of the irreducible representations Γ_i of the point group **G** of the crystal. However, because of the inclusion of spin the representations involved are the double-valued representations of **G** instead of the single-valued representations of **G**.

The simplification of the expansion in eqn. (4.12) by imposing the restriction that it must transform according to a given double-valued representation Γ_i of some point group **G** is very similar to the procedure described in § 4.1 in the absence of spin. Functions which are symmetry-adapted to the double-valued representations Γ_i can be generated by applying the operator W^i defined by eqn. (4.3) to the functions $\alpha Y_l{}^m(\theta, \phi)$ and $\beta Y_l{}^m(\theta, \phi)$. This procedure was applied to all the non-cubic crystallographic point groups by Cracknell (1969 a) and general expressions for all values of l and m were obtained ; those results were reproduced in table 6.8 of Bradley and Cracknell (1972). The same method could also be applied to the cubic point groups, but for the cubic point groups it is not possible to obtain general expressions for all l and m (see, for example, table 12 for some of the single-valued representations). Therefore in practice, instead of applying the operator W^i to $\alpha Y_l^m(\theta, \phi)$ and $\beta Y_l^m(\theta, \phi)$, it was found to be easier to determine symmetry-adapted functions for the double-valued representations of a cubic point group by relating them to the functions that had already been obtained for the single-valued representations of that point group (Cracknell and Joshua 1970). Suppose that Γ_i is one of the single-valued irreducible representations of **G** and that

$$\langle u_p{}^i| = \langle u_1{}^i, u_2{}^i, u_3{}^i, \ldots, u_d{}^i| \tag{4.13}$$

is a basis of Γ_i, where d is the dimension (degeneracy) of Γ_i. Then by studying the function

$$\langle \psi_p{}^k| = \langle \alpha u_1{}^i, \alpha u_2{}^i, \alpha u_3{}^i, \ldots, \alpha u_d{}^i, \beta u_1{}^i, \beta u_2{}^i, \beta u_3{}^i, \ldots, \beta u_d{}^i| \tag{4.14}$$

which forms a basis of the Kronecker product representation

$$\Gamma_k = \Gamma_i \boxtimes \mathscr{D}^{1/2} \tag{4.15}$$

where $\mathscr{D}^{1/2}$ is the representation of **G** to which the spin functions α and β belong, it is possible to determine functions that are symmetry-adapted to the double-valued representations of **G** (see also § 1.5). The results are given in table 6.12 of Bradley and Cracknell (1972) and are reproduced in table 14.

To illustrate the use of table 14 we return to the example of the splitting of an atomic F term in an environment with the symmetry of the point group 432 (O). In the absence of spin effects an atomic F term splits according to

$$\mathscr{D}^3 = \Gamma_2 \oplus \Gamma_4 \oplus \Gamma_5 \tag{2.1}$$
$$(7) \quad (1) \quad (3) \quad (3)$$

(see fig. 7). Since we are still assuming that the atom under consideration has only one outer electron the value of S will be $\frac{1}{2}$ so that we are considering a ^2F term. The splitting of the Γ_2, Γ_4 and Γ_5 terms when spin–orbit coupling is introduced will be different from that illustrated in part (d) of fig. 7 because we are now considering a ^2F term rather than a ^4F term. The spin degeneracy is sometimes also indicated by writing the terms in the crystal as $^2\Gamma_2$, $^2\Gamma_4$ and $^2\Gamma_5$. In 432 (O) $\mathscr{D}^{1/2}$ belongs to Γ_6 (see table 2) and therefore, using table 2, the splitting pattern will be determined by

$$\Gamma_2 \boxtimes \Gamma_6 = \Gamma_7 \tag{4.16}$$

$$\Gamma_4 \boxtimes \Gamma_6 = \Gamma_6 \oplus \Gamma_8 \tag{4.17}$$

$$\Gamma_5 \boxtimes \Gamma_6 = \Gamma_7 \oplus \Gamma_8 \tag{4.18}$$

Table 14.   Symmetry-adapted functions for the cubic double point groups
$(\omega = \exp(2\pi i/3))$

Point group	Representation	Γ^i	Basis
23 (T)	Γ_5	Γ_1	$\langle \alpha u_1{}^i \; ; \; \beta u_1{}^i |$
	Γ_6	Γ_4	$-\alpha u_3{}^i + \beta u_1{}^i - i\beta u_2{}^i \; ; \; \alpha u_1{}^i + i\alpha u_2{}^i + \beta u_3{}^i |$
		Γ_2	$\alpha u_3{}^i \; ; \; \beta u_1{}^i |$
		Γ_4	$\omega \alpha u_3 - \omega^* \beta u_1{}^i + i\beta u_2{}^i \; ; \; -\omega^* \alpha u_1{}^i - i\alpha u_2{}^i - \omega \beta u_3{}^i |$
	Γ_7	Γ_3	$\alpha u_1{}^i \; ; \; \beta u_1{}^i |$
		Γ_4	$-\omega^* \alpha u_3{}^i + \omega \beta u_1{}^i - i\beta u_2{}^i \; ; \; \omega \alpha u_1{}^i + i\alpha u_2{}^i + \omega^* \beta u_3{}^i |$
432 (O)	Γ_6	Γ_1	$\alpha u_1{}^i \; ; \; \beta u_1{}^i |$
		Γ_4	$-\alpha u_3{}^i + \beta u_1{}^i - i\beta u_2{}^i \; ; \; \alpha u_1{}^i + i\alpha u_2{}^i + \beta u_3{}^i |$
	Γ_7	Γ_2	$\alpha u_3{}^i \; ; \; \beta u_1{}^i |$
		Γ_5	$-\alpha u_3{}^i + \beta u_1{}^i - i\beta u_2{}^i \; ; \; \alpha u_1{}^i + i\alpha u_2{}^i + \beta u_3{}^i |$
	Γ_8	Γ_3	$\alpha u_1{}^i \; ; \; \alpha u_2{}^i \; ; \; \beta u_1{}^i \; ; \; \beta u_2{}^i |$
		Γ_4	$-2\alpha u_3{}^i - \beta u_1{}^i + i\beta u_2{}^i \; ; \; \sqrt{3}\beta u_1{}^i + i\sqrt{3}\beta u_2{}^i \; ; \; -\alpha u_1{}^i - i\alpha u_2{}^i + 2\beta u_3{}^i \; ; \; \sqrt{3}\alpha u_1{}^i - i\sqrt{3}\alpha u_2{}^i |$
		Γ_5	$\sqrt{3}\beta u_1{}^i + i\sqrt{3}\beta u_2{}^i \; ; \; 2\alpha u_3{}^i + \beta u_1{}^i - i\beta u_2{}^i \; ; \; \sqrt{3}\alpha u_1{}^i - i\sqrt{3}\alpha u_2{}^i \; ; \; \alpha u_1{}^i + i\alpha u_2{}^i - 2\beta u_3{}^i |$

(Cracknell and Joshua 1970.)

and this is illustrated in fig. 10. Using eqn. (4.7) and table 14 we find that the wave functions for the two states in the Γ_7 term in fig. 10 (b) derived from the $^2\Gamma_2$ term are given by

$$\psi(r, \theta, \phi) = \alpha R_{n3}(r) Y_3{}^{2, \, s}(\theta, \phi) \left.\begin{array}{c}\\[2mm]\end{array}\right\}.$$

$$\beta R_{n3}(r) Y_3{}^{2, \, s}(\theta, \phi) \qquad (4.19)$$

For convenience we abbreviate the expressions for $\psi_{\Gamma_4}(r, \theta, \phi)$ and $\psi_{\Gamma_5}(r, \theta, \phi)$ given in eqns. (4.8) and (4.9) by writing

$$R_{n3}(r)\langle\{a Y_3{}^{1, \, c}(\theta, \phi) - b Y_3{}^{3, \, c}(\theta, \phi)\}, \; \{a Y_3{}^{1, \, s}(\theta, \phi) + b Y_3{}^{3, \, s}(\theta, \phi)\},$$
$$- Y_3{}^0(\theta, \phi)| = \langle u_1{}^4, u_2{}^4, u_3{}^4| \qquad 4.20)$$

and

$$R_{n3}(r)\langle\{- b Y_3{}^{1, \, c}(\theta, \phi) - a Y_3{}^{3, \, c}(\theta, \phi)\}, \; \{b Y_3{}^{1, \, s}(\theta, \phi) - a Y_3{}^{3, \, s}(\theta, \phi)\},$$
$$Y_3{}^{2, \, c}(\theta, \phi)| = \langle u_1{}^5, u_2{}^5, u_3{}^5|. \qquad (4.21)$$

Fig. 10

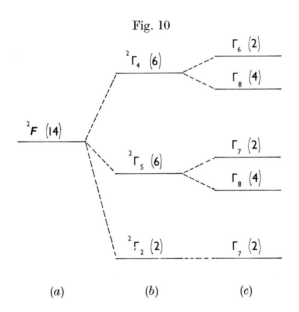

The splitting of an atomic ^2F term in a crystal field with the symmetry of 432 (O), (a) free atom, (b) crystal field alone, (c) strong crystal field and weak spin–orbit coupling.

From table 14 the wave functions of the terms in fig. 10 (b) derived from the $^2\Gamma_4$ and $^2\Gamma_5$ terms are therefore given by

$$\Gamma_6 \quad 3^{-1/2}(-\alpha u_3{}^4 + \beta u_1{}^4 - i\beta u_2{}^4) \qquad \Gamma_7 \quad 3^{-1/2}(-\alpha u_3{}^5 + \beta u_1{}^5 - i\beta u_2{}^5)$$
$$3^{-1/2}(\alpha u_1{}^4 + i\alpha u_2{}^4 + \beta u_3{}^4) \qquad\qquad 3^{-1/2}(\alpha u_1{}^5 + i\alpha u_2{}^5 + \beta u_3{}^5)$$

$$\Gamma_8 \quad 6^{-1/2}(-2\alpha u_3{}^4 - \beta u_1{}^4 + i\beta u_2{}^4) \qquad \Gamma_8 \quad 6^{-1/2}(\sqrt{3}\beta u_1{}^5 + i\sqrt{3}\beta u_2{}^5)$$
$$6^{-1/2}(\sqrt{3}\beta u_1{}^4 + i\sqrt{3}\beta u_2{}^4) \qquad\qquad 6^{-1/2}(2\alpha u_3{}^5 + \beta u_1{}^5 - i\beta u_2{}^5)$$
$$6^{-1/2}(-\alpha u_1{}^4 - i\alpha u_2{}^4 + 2\beta u_3{}^4) \qquad\qquad 6^{-1/2}(\sqrt{3}\alpha u_1{}^5 - i\sqrt{3}\alpha u_2{}^5)$$
$$6^{-1/2}(\sqrt{3}\alpha u_1{}^4 - i\sqrt{3}\alpha u_2{}^4) \qquad\qquad 6^{-1/2}(\alpha u_1{}^5 + i\alpha u_2{}^5 - 2\beta u_3{}^5). \qquad (4.22)$$

In the above account we have obtained symmetry-adapted functions for the double-valued representations of the classical point groups by first introducing a splitting of the energy levels of a free atom or ion by a field with the symmetry of a classical point group and secondly introducing spin effects. That is, in the notation of § 2.1 we have followed steps (A) and (B). An alternative procedure would have been to introduce spin–orbit coupling first and to construct wave functions that form bases of the double-valued representations $\mathscr{D}^j$ ($j =$ half an odd integer) of the rotation group $O(3)$. By studying the splitting of $\mathscr{D}^j$ in a field with the symmetry of a classical point group it would be possible to obtain wave functions for the terms illustrated, for example, in part (c) of fig. 7. This involves using the steps described as (C) and (D) in § 2.1. Since the terms in part (c) of fig. 7 evolve steadily into the terms in part (d) of fig. 7 it is clearly only a matter of convenience, determined by practical considerations, whether one follows steps (A) and (B) or steps (C) and (D) to determine the wave functions for electrons when both spin effects and crystal-field effects are included.

4.3. *Many-electron wave functions*

In considering a term in an atom or ion one should remember that such a term, as determined from spectroscopic measurements for example, is an energy level of the whole atom or ion and therefore, strictly speaking, the wave function for such a term will be a function that involves the coordinates of all the electrons in the atom. An atom or ion can be regarded as consisting of a core, which contains the nucleus and a collection of closed shells of electrons, together with a number of outer electrons in partially filled shells ; fortunately in practice the spectroscopic properties of an atom or ion are determined almost completely by the behaviour of the outer electrons, which are relatively few in number. The account which we have given so far has been restricted to the special case of an atom or ion in which there is only one outer electron.

If the number of outer electrons is larger than 1, the wave function of each of these electrons can be expressed in the form given in eqn. (4.2). It is then necessary to consider how to combine the wave functions for all the outer electrons so as to obtain the complete wave function for the term of the atom or ion. There are two slightly different ways of proceeding. One possibility is first to impose the crystallographic symmetry on the one-electron wave functions of the form of eqn. (4.2) and hence to find the forms of the allowed one-electron wave functions for the atom or ion in a crystal ; that is, we first impose the condition that each of the one-electron wave functions belongs to one of the irreducible representations of the point group **G**. Then by knowing how many electrons are present and allocating these electrons to a certain configuration of the one-electron states it is possible to form products

$$\phi(\mathbf{r}_1, \mathbf{r}_2, \mathbf{r}_3, \ldots) = \phi_i(\mathbf{r}_1)\phi_j(\mathbf{r}_2)\phi_k(\mathbf{r}_3) \ldots \tag{4.23}$$

of the wave functions for those one-electron states that are assumed to be occupied. If the individual functions in eqn. (4.23) were completely arbitrary functions, group theory would not be able to be used very profitably in connection with the wave functions and energy levels of atoms or ions with more than one outer electron. However, each of the one-electron wave functions on the right-hand side of eqn. (4.23) belongs to one of the irreducible representations of **G**; suppose they belong to $\Gamma_i, \Gamma_j, \Gamma_k, \dots$. The symmetry of $\phi(\mathbf{r}_1, \mathbf{r}_2, \mathbf{r}_3, \dots)$ will be that of the Kronecker product $\Gamma_i \boxtimes \Gamma_j \boxtimes \Gamma_k \dots$ of those irreducible representations of **G** to which the one-electron wave functions on the right-hand side of this equation belong. If the representations $\Gamma_i, \Gamma_j, \Gamma_k, \dots$ are degenerate it may be necessary to form linear combinations of the products of one-electron wave functions so as to obtain a function that belongs to (that is, is symmetry-adapted to) one of the irreducible representations of **G**. This can be done by using the tables of coupling constants given by Koster *et al.* (1963); we have already reproduced one extract from these tables in table 3. It must be remembered that the wave function for the whole system must satisfy the principle of anti-symmetry and so it is necessary to anti-symmetrize these product wave functions, or combinations of product wave functions. This can be done by forming Slater determinants.

As an alternative procedure one could first form Slater determinants, using one-electron atomic wave functions, and then subsequently construct linear combinations of these Slater determinants which would be both eigenfunctions of S^2 and also symmetry-adapted to one of the irreducible representations, Γ_i, of **G**. Such a Slater determinant could be written as

$$\psi(\mathbf{r}_1, \mathbf{r}_2, \mathbf{r}_3, \dots) = \begin{vmatrix} \phi_1(\mathbf{r}_1) & \phi_2(\mathbf{r}_1) & \phi_3(\mathbf{r}_1) & \cdots \\ \phi_1(\mathbf{r}_2) & \phi_2(\mathbf{r}_2) & \phi_3(\mathbf{r}_2) & \cdots \\ \phi_1(\mathbf{r}_3) & \phi_2(\mathbf{r}_3) & \phi_3(\mathbf{r}_3) & \cdots \\ \vdots & \vdots & \vdots & \end{vmatrix} \tag{4.24}$$

where $\phi_i(\mathbf{r}_j)$ is a one-electron wave function $\psi_{nl}{}^m(r_j, \theta_j, \phi_j)$ as given in eqn. (4.1) and $\mathbf{r}_j (= (r_j, \theta_j, \phi_j))$ is the position vector of the electron labelled by j. The Slater determinants are necessarily anti-symmetric with respect to the interchange of any pair of electrons. The construction of linear combinations of these determinants to form a basis of Γ_i can then be performed by using, for example, the operator W^i defined by eqn. (4.3).

The above discussion can be adapted to cover the inclusion of spin by using wave functions of the form of eqn. (4.11) in place of wave functions of the form of eqn. (4.1).

We shall illustrate the construction of many-electron wave functions by considering an example. In practice the examples which have been of greatest interest involve the d electrons of transition metal ions in a crystal field where **G** is the cubic point group $m3m$ (O_h); extensive discussions will be found, for example, in the books by Griffith (1961) and Sugano *et al.* (1970). In a crystal field with the symmetry of $m3m$ (O_h) the splitting of the five-fold degenerate d levels can be determined from the compatibility tables between the representations $\mathscr{D}^l$ of $O(3)$ and the irreducible representations of $m3m$ (O_h). From these tables we find

$$\mathscr{D}^2 = \Gamma_3{}^+ \oplus \Gamma_5{}^+ \tag{4.25}$$

where the notation is that of Koster *et al.* (1963) or

$$\mathscr{D}^2 = E_g \oplus T_{2g} \tag{4.26}$$

in the Mulliken notation. If spin is included all these degeneracies will be doubled. Suppose we consider an atom or ion with two d electrons, that is with a d^2 configuration. We include spin in the discussion from the outset so that the one-electron wave functions are

$$\alpha R_{n2}(r) Y_2{}^m(\theta, \phi)$$
$$\beta R_{n2}(r) Y_2{}^m(\theta, \phi)$$

where $m = -2, -1, 0, 1$ or 2. To accommodate two electrons we have to choose any two of these ten available states, subject to the restriction imposed by the Pauli exclusion principle, or the principle of anti-symmetry, that no two electrons may be in the same state. The number of different ways of choosing two functions from ten functions is $(10!/8!2!) = 45$.

In a cubic field it is convenient to separate the 45 possible two-electron wave functions into three sets :

$t_{2g}{}^2$ when both the one-electron wave functions belong to the T_{2g} set of functions $(6!/4!2! = 15)$,

$e_g{}^2$ when both the one-electron wave functions belong to the E_g set of functions $(4!/2!2! = 6)$,

$t_{2g}e_g$ when one of the one-electron wave functions belongs to T_{2g} and the other belongs to E_g $(6 \times 4 = 24)$.

The numbers in brackets indicate the number of these two-electron states in each set. In the cubic field the one-electron wave functions belonging to T_{2g} and E_g become (see, for example, table 2.6 of Bradley and Cracknell (1972)) :

$$
\begin{aligned}
T_{2g} \quad & \alpha R_{n2}(r)\{Y_2{}^1(\theta, \phi) + Y_2{}^{-1}(\theta, \phi)\}/\sqrt{2} \quad && \xi \\[4pt]
& \beta R_{n2}(r)\{Y_2{}^1(\theta, \phi) + Y_2{}^{-1}(\theta, \phi)\}/\sqrt{2} \quad && \bar{\xi} \\[4pt]
& \alpha R_{n2}(r)\{Y_2{}^1(\theta, \phi) - Y_2{}^{-1}(\theta, \phi)\}/\sqrt{2} \quad && \eta \\[4pt]
& \beta R_{n2}(r)\{Y_2{}^1(\theta, \phi) - Y_2{}^{-1}(\theta, \phi)\}/\sqrt{2} \quad && \bar{\eta} \\[4pt]
& \alpha R_{n2}(r)\{Y_2{}^2(\theta, \phi) - Y_2{}^{-2}(\theta, \phi)\}/\sqrt{2} \quad && \zeta \\[4pt]
& \beta R_{n2}(r)\{Y_2{}^2(\theta, \phi) - Y_2{}^{-2}(\theta, \phi)\}/\sqrt{2} \quad && \bar{\zeta} \\[4pt]
E_g \quad & \alpha R_{n2}(r) Y_2{}^0(\theta, \phi) \quad && u \\[4pt]
& \beta R_{n2}(r) Y_2{}^0(\theta, \phi) \quad && \bar{u} \\[4pt]
& \alpha R_{n2}(r)\{Y_2{}^2(\theta, \phi) + Y_2{}^{-2}(\theta, \phi)\}/\sqrt{2} \quad && v \\[4pt]
& \beta R_{n2}(r)\{Y_2{}^2(\theta, \phi) + Y_2{}^{-2}(\theta, \phi)\}/\sqrt{2} \quad && \bar{v}
\end{aligned}
\tag{4.27}
$$

where factors like $-i$ have been ignored. For convenience the symbols on the extreme right will be used to represent these functions.

Suppose we consider the $t_{2g}{}^2$ configuration, then we shall have 15 product wave functions such as

$$\xi(1)\eta(2) \tag{4.28}$$

where the number in brackets is used to distinguish the two electrons. Anti-symmetrizing these products will give 15 Slater determinants such as

$$
\begin{vmatrix} \xi(1) & \eta(1) \\ \xi(2) & \eta(2) \end{vmatrix}. \tag{4.29}
$$

It is convenient to introduce an even more condensed notation for these Slater determinants, for example

$$
|\xi \quad \eta| \tag{4.30}
$$

for the determinant in eqn. (4.29). The eigenvalue of S_z for each Slater determinant can be seen by inspection; thus $|\xi\eta|$ has two α functions and so $S_z = +1$, for $|\xi\bar{\eta}|$ there is one α and one β function so that $S_z = 0$ and for $|\bar{\xi}\bar{\eta}|$ there are two β functions so that $S_z = -1$.

To obtain wave functions which belong to one of the irreducible representations of $m3m$ (O_h) it may be necessary to take linear combinations of the Slater determinants. These linear combinations can be obtained either by inspection or by using the operator W^i defined in eqn. (4.3), and the results are given in table 15. If one simply wishes to know which irreducible representations are allowed for the t_{2g}^2 wave functions one can examine the reduction of the Kronecker product $T_{2g} \boxtimes T_{2g}$ (or $\Gamma_5^+ \boxtimes \Gamma_5^+$), see table 2,

$$
\Gamma_5^+ \boxtimes \Gamma_5^+ = \Gamma_1^+ \oplus \Gamma_3^+ \oplus \Gamma_4^+ \oplus \Gamma_5^+ \tag{4.31}
$$

or

$$
T_{2g} \boxtimes T_{2g} = A_{1g} \oplus E_g \oplus T_{1g} \oplus T_{2g}. \tag{4.32}
$$

Table 15. t_{2g}^2 wavefunctions

Repre-sentation	Wave function	S_z
$^1A_{1g}$	$\|\xi\bar{\xi}\| + \|\eta\bar{\eta}\| + \|\zeta\bar{\zeta}\|$	0
^1E_g	$-\|\xi\bar{\xi}\| - \|\eta\bar{\eta}\| + 2\|\zeta\bar{\zeta}\|$	0
	$\|\xi\bar{\xi}\| - \|\eta\bar{\eta}\|$	0
$^3T_{1g}$	$\|\xi\eta\|$	1
	$\|\eta\zeta\|$	1
	$\|\zeta\xi\|$	1
	$\|\xi\bar{\eta}\| - \|\eta\bar{\xi}\|$	0
	$\|\eta\bar{\zeta}\| - \|\zeta\bar{\eta}\|$	0
	$\|\zeta\bar{\xi}\| - \|\xi\bar{\zeta}\|$	0
	$\|\bar{\xi}\bar{\eta}\|$	-1
	$\|\bar{\eta}\bar{\zeta}\|$	-1
	$\|\bar{\zeta}\bar{\xi}\|$	-1
$^1T_{2g}$	$\|\xi\bar{\eta}\| + \|\eta\bar{\xi}\|$	0
	$\|\eta\bar{\zeta}\| + \|\zeta\bar{\eta}\|$	0
	$\|\zeta\bar{\xi}\| + \|\xi\bar{\zeta}\|$	0

Note. These functions are not necessarily normalized. (Sugano *et al.* 1970.)

The usefulness of this is rather restricted. Thus it indicates that there is no t_{2g}^2 two-electron wave function belonging to A_{2g}. However, because it does not allow for the inclusion of spin it does not indicate the spin degeneracies of the t_{2g}^2 states; that is, it does not show that the T_{1g} functions have a three-fold spin degeneracy ($S=1$, $S_z=1, 0, -1$) in addition to the three-fold orbital degeneracy, whereas the T_{2g} functions do not have this spin degeneracy (see table 15). Neither can this information be obtained from the reduction of the Kronecker product of the double-valued representations $(\Gamma_7 \oplus \Gamma_8)$ of $m3m$ (O_h) to which the functions ξ, η, ζ, $\bar{\xi}$, $\bar{\eta}$ and $\bar{\zeta}$ (or linear combinations thereof) belong. The reason for this is that the Kronecker product $(\Gamma_7 \oplus \Gamma_8) \boxtimes (\Gamma_7 \oplus \Gamma_8)$ would also contain representations corresponding to product functions such as $\xi\xi$, etc. which are not allowed by the Pauli exclusion principle.

We have only given the details of the construction of the two-electron wave functions for the t_{2g}^2 configuration. The arguments can be repeated and tables similar to table 15 can be constructed for the e_g^2 and $t_{2g}e_g$ configurations. We shall not give all the details here; the results are given in tables 2.3 and 2.4 of Sugano et al. (1970).

This example was only for a two-electron configuration. The argument can be extended to a configuration involving a greater number of electrons. The details of some examples involving more than two d electrons in a cubic crystal field are given at the beginning of chapter 3 of the book by Sugano et al. (1970).

4.4. *The calculation of crystal-field splittings*

In §§ 4.1 and 4.2 we were concerned with exploiting the symmetry of the crystal field to simplify the form of the wave function in eqn. (4.2) or eqn. (4.12) for an atom or ion with one outer electron and in § 4.3 we have considered the extension of these ideas to an atom or ion with more than one outer electron. Having achieved the maximum possible simplification of the wave function it remains to calculate the actual splittings of the energy levels of the atom or ion in the crystal. The wave functions that we have obtained in various equations have no remaining unknown coefficients and so all that remains to be done is to assume some form for the electrostatic potential and to calculate the energy levels in the crystal, using perturbation theory. Suppose that $V(r, \theta, \phi)$ is the difference between the potential experienced by an electron in an atom or ion in a crystal and the potential experienced by that electron in the same atom or ion in free space. If $V(r, \theta, \phi)$ is known, then the wave functions for that electron in a crystal, such as the functions in eqns. (4.7) and (4.9), can be used to calculate the energy levels of that electron in the crystal. However, in practice one is unlikely to be able to calculate $V(r, \theta, \phi)$ accurately from first principles. Remembering that $V(r, \theta, \phi)$ must be a solution of Poisson's equation $\nabla^2 V = -\rho/\epsilon_0$ we could expand it in the form

$$V(r, \theta, \phi) = \sum_{l, m} A_l^m(r) Y_l^m(\theta, \phi) \tag{4.33}$$

where the $A_l^m(r)$ are functions of r. If we neglect the effect of the charge density in the ligand or bond itself, $V(r, \theta, \phi)$ would satisfy Laplace's equation rather than Poisson's equation and $A_l^m(r)$ could be written as $B_l^m r^l$ where the B_l^m are constants; then

$$V(r, \theta, \phi) = \sum_{l, m} B_l^m r^l Y_l^m(\theta, \phi). \qquad (4.34)$$

We shall use eqn. (4.34) although it should, however, be stressed that this is only an approximation to eqn. (4.33). The matrix elements take the form

$$\iiint \psi^*(r, \theta, \phi) V(r, \theta, \phi) \psi(r, \theta, \phi) r^2 \sin \theta \, dr \, d\theta \, d\phi$$

$$= \iiint \{ \sum_{n_1 l_1 m_1} A_{n_1 l_1}^{m_1} R_{n_1 l_1}(r) Y_{l_1}^{m_1}(\theta, \phi) \}^*$$

$$\times \{ \sum_{l_2 m_2} B_{l_2}^{m_2} r^{l_2} Y_{l_2}^{m_2}(\theta, \phi) \} \{ \sum_{n_3 l_3 m_3} A_{n_3 l_3}^{m_3} R_{n_3 l_3}(r) Y_{l_3}^{m_3}(\theta, \phi) \}$$

$$\times r^2 \sin \theta \, dr \, d\theta \, d\phi \qquad (4.35)$$

where the coefficients A_{nl}^m are known (as in eqn. (4.7) or in eqn. (4.9) for example). There are several advantages to be gained from using an expansion of $V(r, \theta, \phi)$ in terms of spherical harmonics. The first is that by using the known properties of the spherical harmonics the integrations over θ and ϕ in eqn. (4.35) can be obtained analytically (for some examples see Stevens (1952), Elliott and Stevens (1953)). The integral over r in eqn. (4.35) can also be calculated so that this matrix element can be evaluated completely, apart from the coefficients B_l^m. These coefficients B_l^m can therefore be regarded as parameters that have to be determined by evaluating the matrix elements in terms of these coefficients, and then fitting the matrix elements to the splitting pattern which has been determined experimentally from spectroscopic measurements. The other advantages of using an expansion of $V(r, \theta, \phi)$ in terms of spherical harmonics arise from the simplifications which can be achieved in the form of $V(r, \theta, \phi)$ because of the point-group symmetry that $V(r, \theta, \phi)$ possesses.

The fact that $V(r, \theta, \phi)$ is taken to have the symmetry of a certain point group **G** means that we can simplify eqn. (4.34) by retaining only those terms that are invariant under the symmetry operations of **G**. Expressed in terms of representation theory this means that we only retain those terms involving spherical harmonics that belong to the identity representation Γ_1 (or Γ_1^+) of **G**. In the two examples which we considered in § 4.1 we deliberately used the same atomic term, namely an F term, in each one so that we can also indicate the importance of the concept of hierarchies of symmetry in this work. By this we mean that the point-group symmetry that one ascribes to the environment of an atom or ion in a crystal may depend on how carefully one studies the system. Thus for a cubic crystal field belonging to 432 many of the terms in eqn. (4.34) vanish; the only terms that survive (up to $l = 6$) would be

$$V_c(r, \theta, \phi) = a_0 + a_4 r^4 \{ c Y_4^{0, c}(\theta, \phi) + d Y_4^{4, c}(\theta, \phi) \}$$

$$+ a_6 r^6 \{ e Y_6^{0, c}(\theta, \phi) + f Y_6^{4, c}(\theta, \phi) \} \qquad (4.36)$$

where the values of the coefficients c, d, e and f are given in the entries for $\Gamma_1(A_1)$ of 432 in table 12. The lower the order of the point group of the symmetry of the crystal field, the larger will be the number of terms that it will be necessary to retain in an expansion of the form of eqn. (4.34) up to a given order of l.

The calculation of matrix elements, in terms of the coefficients in the expansion of $V(r, \theta, \phi)$, is basically a computational problem. There are two ways in which group theory may be able to simplify this problem; one is by showing that certain matrix elements must vanish as a result of symmetry considerations and the other is by obtaining linear relationships among sets of different matrix elements. In either case, the effect is to simplify the computations by reducing the number of independent and non-vanishing

Fig. 11

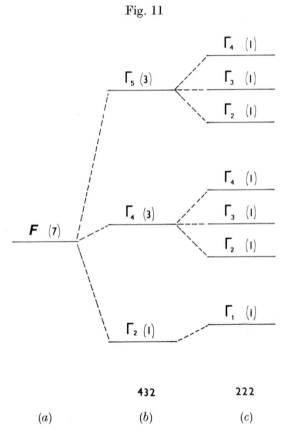

$$432 \qquad\qquad 222$$

$$(a) \qquad\qquad (b) \qquad\qquad (c)$$

Atomic F term with a weak orthorhombic field added to a strong cubic field.

matrix elements, or contributions to matrix elements, that actually have to be evaluated by integration. The demonstration that certain matrix elements must vanish involves using Neumann's principle. Consider the matrix element $\langle \psi_i V \psi_j \rangle$, where ψ_i, V and ψ_j belong to Γ_i, Γ_V and Γ_j, respectively, of **G**. This matrix element must vanish if the triple Kronecker product $\Gamma_i^* \boxtimes \Gamma_V \boxtimes \Gamma_j$ does not contain the identity representation of **G**. We shall illustrate the vanishing of certain matrix elements in the following few paragraphs. The determination of relationships between different matrix elements, or between different contributions to a given matrix element, arises from the relationships which exist between the values of the integral

$$\int\int Y_{l_1}{}^{m_1}(\theta, \phi)\ Y_{l_2}{}^{m_2}(\theta, \phi)\ Y_{l_3}{}^{m_3}(\theta, \phi)\ \sin\theta\ d\theta\ d\phi$$

for a given set of values of l_1, l_2 and l_3 but with different sets of values of m_1, m_2 and m_3. This is based on the use of the Wigner–Eckart theorem (for further details see one of the recent specialized books such as those by Abragam and Bleaney (1970) and Sugano *et al.* (1970)).

It happens quite frequently that the electrostatic field experienced by an atom or ion in a crystal can be described to a good first approximation by a point group of rather high symmetry, such as 432 (O) for example, but that a more refined study shows that the electrostatic field is actually of rather low symmetry, such as 222 (D_2) for example. The effect of a strong cubic field with a weak orthorhombic field superimposed is illustrated in fig. 11, where we have made use of the compatibilities between 432 and 222 (see table 13). In this case one would first use the measured positions of the levels in fig. 11 (*b*) to determine the parameters for the cubic field by fitting to the matrix elements obtained using the cubic field as a perturbation, with the atomic wave functions in eqn. (4.4) as the unperturbed wave functions and the cubic wave functions given in eqn. (4.9) as the perturbed wave functions. Having determined the parameters in the expansion of the cubic field in eqn. (4.36), one could then consider the additional terms which would be present when eqn. (4.34) is simplified using the orthorhombic point group 222 instead of the cubic point group 432. The coefficients of these additional terms can then be determined by calculating the matrix elements, using the orthorhombic field as a perturbation with the cubic wave functions in eqn. (4.9) as the unperturbed wave functions, and the orthorhombic wave functions in eqn. (4.7) as the perturbed wave functions, and then fitting the matrix elements to the measured positions of the levels in fig. 11 (*c*). From the entry for $\Gamma_1(A_1)$ in table 11 we see that there are many additional terms in the orthorhombic field ; only going as far as $l=4$ we have

$$V_0(r,\ \theta,\ \phi) = a_0 + a_2{}^0 r^2 Y_2{}^{0,\ c}(\theta,\ \phi) + a_2{}^2 r^2 Y_2{}^{2,\ c}(\theta,\ \phi)$$

$$+ a_3{}^2 r^3 Y_3{}^{2,\ s}(\theta,\ \phi) + a_4{}^0 r^4 Y_4{}^{0,\ c}(\theta,\ \phi) + a_4{}^2 r^4 Y_4{}^{2,\ c}(\theta,\ \phi)$$

$$+ a_4{}^4 r^4 Y_4{}^{4,\ c}(\theta,\ \phi). \qquad (4.37)$$

The positions at which one chooses to terminate expansions for $V(r,\ \theta,\ \phi)$, such as those in eqns. (4.36) and (4.37), will be determined by the number of pieces of experimental data available for the splitting pattern in question.

In terms of representation theory, each term on the right-hand side of the expression for the cubic potential in eqn. (4.36) belongs to the identity representation Γ_1 of the cubic point group 432. Similarly, each term on the right-hand side of eqn. (4.37) belongs to the identity representation Γ_1 of 222. However, 222 is a subgroup of 432 and we could re-arrange the terms on the right-hand side of eqn. (4.37) so that they could be identified as bases of various representations of 432. This requires the use of tables of basis functions for the cubic point groups ; from the entries for 432 (O) in table 12 and

Table 16.   Symmetry-adapted functions for the two-dimensional representations of the cubic point groups

43m (T_d)	432 (O)	m3m (O_h)	l	φ-dep	Function
E	E	E_g	2	c	1(0)
				c	1(2)
E	E	E_g	4	c	0·64549722437(0) − 0·76376261583(4)
				c	−1(2)
E*	E	E_u	5	s	1(4)
				s	−1(2)
E	E	E_g	6	c	0·93541434670(0) + 0·35355339059(4)
				c	0·55901699438(2) + 0·82915619759(6)
E*	E	E_u	7	s	1(4)
				s	0·67700320039(2) − 0·73598007219(6)
E	E	E_g	8	c	0·69597054536(0) − 0·39400753227(4) − 0·60031913556(8)
				c	−0·65068202432(2) − 0·75935031654(6)
E	E	E_g	8	c	0·83601718355(4) − 0·54870326117(8)
				c	−0·75935031654(2) + 0·65068202432(6)
E*	E	E_u	9	s	0·54006172487(4) + 0·84162541153(8)
				s	−0·90138781887(2) − 0·43301270189(6)
E	E	E_g	10	c	0·91144345226(0) + 0·26465657643(4) + 0·31500433312(8)
				c	0·44497917425(2) + 0·48620517700(6) + 0·75206253752(10)
E	E	E_g	10	c	0·76564149349(4) − 0·64326752090(8)
				c	0·39845246619(2) − 0·85957253477(6) + 0·31995419931(10)
E*	E	E_u	11	s	0·49530506035(4) + 0·86871911295(8)
				s	0·74651970280(2) − 0·40934969512(6) − 0·52453899801(10)
E*	E	E_u	11	s	0·86871911295(4) − 0·49530506035(8)
				s	0·78834974923(6) − 0·61522733432(10)
E	E	E_g	12	c	0·71852351504(0) − 0·30406126533(4) − 0·33728552687(8) − 0·52679139953(12)
				c	−0·51600907827(2) − 0·54714937497(6) − 0·65906160002(10)
E	E	E_g	12	c	0·70456648687(4) + 0·33875374295(8) − 0·62356392392(12)
				c	−0·83033956777(2) + 0·13051364479(6) + 0·54175860926(10)

Notes. For the representations marked with an asterisk the partners must be interchanged and the sign of one of them reversed. For 432 (O) E is an alternative label for Γ_3 which was used in table 2. The functions are given as linear combinations of $Y_l^{m,c}(\theta,\phi)$ or of $Y_l^{m,s}(\theta,\phi)$, where the appropriate value of m is given in brackets following each coefficient in the table. For further details see chapter 2 of Bradley and Cracknell (1972).

in table 16 we obtain

$$V_0(r, \theta, \phi) = a_0 \qquad\qquad \Gamma_1$$

$$+ a_4 r^4 \{c\, Y_4^{0,\ c}(\theta, \phi) + d\, Y_4^{4,\ c}(\theta, \phi)\} \qquad \Gamma_1$$

$$\left.\begin{aligned} &+ b_4 r^4 \{d\, Y_4^{0,\ c}(\theta, \phi) - c\, Y_4^{4,\ c}(\theta, \phi)\} \\[2mm] &+ a_4'^4 r^4 Y_4^{2,\ c}(\theta, \phi) \end{aligned}\right\} \quad \Gamma_3$$

$$\left.\begin{aligned} &+ a_2^0 r^2 Y_2^0(\theta, \phi) \\[2mm] &+ a_2^2 r^2 Y_2^{2,\ c}(\theta, \phi) \end{aligned}\right\} \quad \Gamma_3$$

$$+ a_3^2 r^3 Y_3^{2,\ s}(\theta, \phi) \qquad\qquad \Gamma_2 \qquad (4.38)$$

where we have identified the representation of 432 to which each term belongs. Suppose we use a cubic potential $V_c(r, \theta, \phi)$, which belongs to Γ_1 of 432, with a wave function which belongs to some representation Γ_i of 432. Then the matrix element $\langle \psi_i V \psi_i \rangle$ belongs to $\Gamma_i^* \boxtimes \Gamma_1 \boxtimes \Gamma_i$. The reduction of this triple Kronecker product will necessarily contain Γ_1 of 432, irrespective of Γ_i. Therefore, invoking Neumann's principle, for any Γ_i there is no group-theoretical restriction that would require $\langle \psi_i V \psi_i \rangle$ to vanish. Similarly, if we use an orthorhombic potential belonging to 222 with wave functions that belong to bases of 222 there will also be no group-theoretical requirement that $\langle \psi_i V \psi_i \rangle$ vanishes. However, if we use the orthorhombic potential $V_0(r, \theta, \phi)$ with the cubic wave functions (regarded as the unperturbed wave functions) then it may be possible to show that the contributions to $\langle \psi_i V \psi_i \rangle$ from some of those additional terms that are in $V_0(r, \theta, \phi)$ but are not in $V_c(r, \theta, \phi)$ necessarily vanish. For example, suppose we consider the contribution to $\langle \psi_{\Gamma_5} V_0 \psi_{\Gamma_5} \rangle$ arising from a term in $V_0(r, \theta, \phi)$ that belongs to Γ_j of 432. This contribution will belong to $\Gamma_5^* \boxtimes \Gamma_j \boxtimes \Gamma_5$; Γ_5^* is equivalent to Γ_5 and from table 2 we see that

$$\Gamma_5 \boxtimes \Gamma_5 = \Gamma_1 \oplus \Gamma_3 \oplus \Gamma_4 \oplus \Gamma_5. \qquad (4.39)$$

Therefore, in order that the reduction of $\Gamma_5 \boxtimes \Gamma_j \boxtimes \Gamma_5$ contains Γ_1 the representation Γ_j must be Γ_1, Γ_3, Γ_4 or Γ_5. In other words, since the term $a_3^2 r^3 Y_3^{2,\ s}(\theta, \phi)$ in $V_0(r, \theta, \phi)$ in eqn. (4.38) belongs to Γ_2 this term's contribution must vanish. The knowledge, from the group-theoretical arguments, that certain contributions must vanish saves one the trouble of actually evaluating those contributions numerically.

4.5. *The effect of magnetic ordering*

We now have to consider the adaptation of all that we have said so far about crystal-field theory, or ligand-field theory, to the situation in which the symmetry of the environment of an atom or ion in a crystal is described not by one of the 32 classical point groups but by one of the type III Heesch–Shubnikov point groups (black and white point groups) **M**. Suppose that the symmetry of the crystal field experienced by an ion in a crystal is described by one of the magnetic point groups, **M**, given by

$$\mathbf{M} = \mathbf{H} + A\,\mathbf{H}. \qquad (1.11)$$

We then have to consider the modifications that this introduces both to the theory of the splitting pattern of the atomic energy levels (§ 2.1) and also to the forms of the one-electron wave functions, both in the absence of spin (§ 4.1) and when spin is included (§ 4.2). In the conventional sense the term 'crystal field' or 'ligand field' is understood to be the electrostatic field experienced by an atom or ion in a crystal. However, in the case of a magnetic crystal we extend the meaning to include the very large magnetic exchange and anisotropy fields which exist.

Suppose that $RO(3)$ is the direct product of $O(3)$, the rotation group in three dimensions, with the group consisting of the elements E and θ. Suppose also that $D\mathscr{D}^L$, $D\mathscr{D}^S$ and $D\mathscr{D}^J$ are the irreducible co-representations of $RO(3)$ derived from the irreducible representations $\mathscr{D}^L$, $\mathscr{D}^S$ and $\mathscr{D}^J$ of $O(3)$. Then the procedure which was described in § 2.1 for non-magnetic crystals can readily be adapted to the case of magnetic crystals. The steps that we have called (A), (B), (C) and (D) in § 2.1 become

(A) (i)→(v), subduction of the single-valued irreducible co-representations of $RO(3)$ on to the single-valued irreducible co-representations of **M**.

(B) (v)→(iv), reduction of the Kronecker product of the co-representations of **M** obtained in (A) and the co-representations $D\mathscr{D}^S$.

(C) (i)→(ii), reduction of the Kronecker products of the irreducible co-representations $D\mathscr{D}^L$ and $D\mathscr{D}^S$ of $RO(3)$ corresponding to the orbital and spin angular momenta.

(D) (ii)→(iii), subduction of the (single-valued or double-valued) irreducible co-representation $D\mathscr{D}^J$ of $RO(3)$ on to the irreducible co-representations of **M**.

These four steps resemble quite closely the corresponding steps for non-magnetic crystals. The necessary manipulations of the subduction of representations or of the reduction of Kronecker products of representations for non-magnetic crystals can all be achieved by using the characters of the matrices in the representations. The usefulness of the characters of a representation is associated with the fact that the character of the matrix representative $\mathbf{D}(R_j)$ of an element R_j in a given representation is invariant under a unitary transformation

$$\mathbf{D}'(R_j) = \mathbf{S}^{-1}\mathbf{D}(R_j)\mathbf{S} \tag{4.40}$$

where **S** is any unitary matrix. For magnetic crystals, however, the characters of the matrices that form a co-representation have to be treated with care. This is because of complications connected with the definition of equivalence for co-representations; two co-representations are said to be equivalent if there exists a unitary matrix **S** such that eqn. (4.40) holds for all the unitary elements R_j and

$$\mathbf{D}'(A_j) = \mathbf{S}^{-1}\mathbf{D}(A_j)\mathbf{S}^* \tag{4.41}$$

for all the anti-unitary elements A_j. Whereas the character of a matrix corresponding to a unitary element R_j is invariant under the transformation in eqn. (4.40), the character of an anti-unitary element A_j is not necessarily invariant under the transformation in eqn. (4.41). Thus the characters of the matrices for the anti-unitary elements are of relatively little use. However, since any irreducible co-representation of **M** is completely and uniquely specified by one of the irreducible representations of **H**, its unitary subgroup, we can perform the required subduction or reduction for the irreducible representations of **H** and then identify the result in terms of the irreducible co-representations of **M**. It is important not to forget the last step, because although the irreducible co-representations of **M** are unambiguously related to the irreducible representations of **H** the degeneracies of the irreducible co-representations of **M** may not be the same as the degeneracies of the irreducible representations of **H**.

Steps (A) and (D) involve the subduction of one of the irreducible co-representations $D\mathscr{D}^L$ or $D\mathscr{D}^J$ of $RO(3)$ on to the magnetic point group **M**. To do this directly would necessitate using the complete matrix representatives for the co-representations of $RO(3)$ and of **M**; this would make no use of the simplification afforded by the use of characters. The following indirect route from $RO(3)$ to $O(3)$, to **H**, and then to **M** is much easier to use in practice :

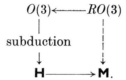

The subduction from $O(3)$ to **H** can be obtained from the tables given by Koster *et al.* (1963) and reproduced in tables 2.7 and 6.6 of Bradley and Cracknell (1972). The determination of the relationship between the irreducible representations of **H** and the irreducible co-representations of **M** can be achieved fairly readily (Dimmock and Wheeler 1962 b, Cracknell 1966 a, b, Cracknell and Wong 1967, Bradley and Cracknell 1972). Complete tables giving the results of the subduction from $RO(3)$ to **M** are given by Cracknell (1968 b). For (A), L is an integer so that $D\mathscr{D}^L$ is one of the single-valued co-representations of $RO(3)$. For (D), J may be an integer or half an odd integer ; if J is an integer $D\mathscr{D}^J$ will also be a single-valued co-representation of $RO(3)$ but if J is half an odd integer $D\mathscr{D}^J$ will be a double-valued co-representation of $RO(3)$.

Steps (B) and (C) involve the reduction of the Kronecker products of co-representations ; in (C) it is the Kronecker product $D\mathscr{D}^L \boxtimes D\mathscr{D}^S$ of co-representations of $RO(3)$ and in (B) it is the Kronecker product of co-representations of **M**. In each case we examine the corresponding Kronecker product of the related representations of the unitary subgroup. For (C) this is easy because

$$\mathscr{D}^L \boxtimes \mathscr{D}^S = \mathscr{D}^{L+S} \oplus \mathscr{D}^{L+S-1} \oplus \ldots \oplus \mathscr{D}^{|L-S|} \qquad (4.42)$$

and for (B) the reductions are given for all possible **H** in the tables of Koster *et al.* (1963). The reductions of the corresponding Kronecker products in $RO(3)$ or in **M** can then be found by using the relationship between the irreducible co-representation of a non-unitary group and the irreducible representations of its unitary subgroup.

As an example we consider a ^2D term in a crystal field with the symmetry of the magnetic point group $4'/mmm'$. The irreducible co-representations of $4'/mmm'$ are identified, in relation to the irreducible representations of the unitary subgroup mmm, in table 17 ; the reductions of their Kronecker

Fig. 12

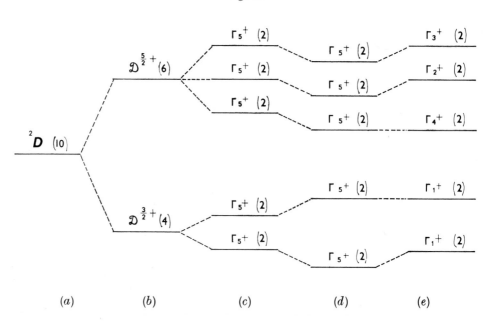

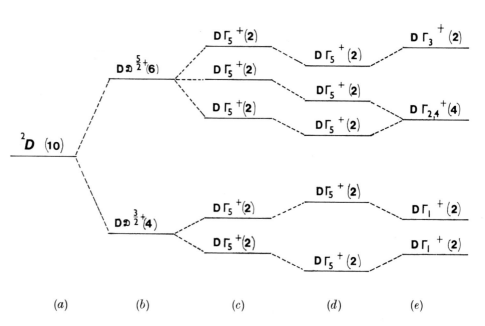

Possible splitting of an atomic ^2D term in mmm, and $4'/mmm'$. (The connectivities between (b) and (e) are not uniquely determined.)

Table 17. Irreducible co-representations of the point group $4'22'$

(a) Character table

	E	$\bar{E}$	$C_{2x}, \bar{C}_{2x}$	$C_{2y}, \bar{C}_{2y}$	$C_{2z}, \bar{C}_{2z}$	
Γ_1	1	1	1	1	1	a
Γ_3	1	1	-1	-1	1	a
Γ_2	1	1	-1	1	-1	$\left.\begin{array}{c} \\ \\ \end{array}\right\} c$
Γ_4	1	1	1	-1	-1	
Γ_5	2	-2	0	0	0	a

The characters are those of the irreducible representations of the unitary sub-group 222 (D_2) ; in the last column the type of the co-representation of $4'22'$ derived from each of these representations is indicated.

(b) Reductions of Kronecker products of irreducible representations of the unitary subgroup

Γ_1	Γ_2	Γ_3	Γ_4	Γ_5	
Γ_1	Γ_2	Γ_3	Γ_4	Γ_5	Γ_1
	Γ_1	Γ_4	Γ_3	Γ_5	Γ_2
		Γ_1	Γ_2	Γ_5	Γ_3
			Γ_1	Γ_5	Γ_4
				$\Gamma_1\oplus\Gamma_2\oplus\Gamma_3\oplus\Gamma_4$	Γ_5

Note. $4'/mmm' = 4'22' \otimes \bar{\mathrm{I}}$.

Table 18. Compatibilities between co-representations of $4'22'$ and of the three-dimensional rotation group

$\mathrm{D}\mathscr{D}^0$	$\mathrm{D}\Gamma_1$ 1(1)	$\mathscr{D}^{2m+\lambda} = m \operatorname{reg} \oplus \mathscr{D}^\lambda$
$\mathrm{D}\mathscr{D}^1$	$\mathrm{D}\Gamma_3 \oplus \mathrm{D}\Gamma_{2,4}$ 1(1) 1(2)	
$\mathrm{D}\mathscr{D}^j$	$(j+\frac{1}{2})\mathrm{D}\Gamma_5$ $(j+\frac{1}{2})(2)$	$j =$ half-odd-integer

products are also given in this table, while the compatibilities with the co-representations of $RO(3)$ are given in table 18. The easiest way to proceed is perhaps first to consider all the possible relative strengths of the spin–orbit coupling and the crystal field in a situation with the symmetry of the unitary subgroup, mmm. This can be done using the same approach that we used in § 2.1 for a ^4F term in 432, see fig. 7. Without describing all the details we simply present the result in fig. 12. From table 17 we see that Γ_1^+, Γ_3^+ and Γ_5^+ of mmm lead to case (a) co-representations of $4'/mmm'$ while Γ_2^+ and Γ_4^+ stick together in a case (c) co-representation of $4'/mmm'$. The splitting pattern of a ^2D atomic term in a crystal field with the symmetry of $4'/mmm'$, for various relative strengths of the spin–orbit coupling and of the crystal field, is also illustrated in fig. 12.

We turn now to the problem of the determination of the forms of the one-electron wave functions for a crystal field with the symmetry of one of the magnetic point groups

$$\mathbf{M} = \mathbf{H} + A\,\mathbf{H}. \tag{1.11}$$

Once again it is most convenient to proceed via the unitary subgroup $\mathbf{H}$. Suppose that the approach described in §§ 4.1 and 4.2 has been used with the symmetry operations of $\mathbf{H}$, the unitary subgroup of $\mathbf{M}$, to determine wave functions that are symmetry-adapted to the various irreducible representations Γ_i of $\mathbf{H}$. Γ_i may be either a single-valued or a double-valued representation. It is then a fairly simple matter to obtain the functions that belong to the irreducible co-representations $D\Gamma_i$ of $\mathbf{M}$ from the functions that belong to Γ_i of $\mathbf{H}$. This can be done by studying the effect of the anti-unitary elements $A\,\mathbf{H}$ on these functions. This in turn can be simplified to the problem of studying the effect of the anti-unitary operation A on these wave functions, where A can be regarded as the additional generating element, or augmenter, that generates $\mathbf{M}$ from $\mathbf{H}$. Suppose we consider the wave functions which form components of the vector function $\langle u_p{}^i |$ which is a basis of Γ_i of $\mathbf{H}$. Then the function $\langle u'_p{}^i |$ defined by

$$\langle u'_p{}^i | = A \langle u_p{}^i | \tag{4.43}$$

forms a basis of the irreducible representation $\bar{\Gamma}_i$ of $\mathbf{H}$. If $D\Gamma_i$ belongs to case (a) $\langle u'_p{}^i |$ and $\langle u_p{}^i |$ are not linearly independent and the degeneracy of $D\Gamma_i$ is the same as the degeneracy of Γ_i. Consequently in case (a) $\langle u_p{}^i |$ can be taken as a basis of Γ_i so that the wave functions for the term $D\Gamma_i$ in $\mathbf{M}$ are the same as for the term Γ_i in $\mathbf{H}$. In case (b) the term $D\Gamma_i$ in $\mathbf{M}$ arises by the sticking together of two terms that in $\mathbf{H}$ were not required by symmetry to be degenerate, but which had the same transformation properties under the operations of $\mathbf{H}$, that is they both belong to the same irreducible representation, Γ_i of $\mathbf{H}$. Although $\langle u'_p{}^i |$ is linearly independent of $\langle u_p{}^i |$ it can still be generated from $\langle u_p{}^i |$ by using eqn. (4.43). If $D\Gamma_i$ belongs to case (c) $\langle u'_p{}^i |$ and $\langle u_p{}^i |$ again are independent functions but this time they belong to two different irreducible representations Γ_i and Γ_j $(= \bar{\Gamma}_i)$ of $\mathbf{H}$. In case (c) the term $D\Gamma_{i,\,j}$ (or $D\Gamma_{j,\,i}$) therefore arises by the sticking together of two terms that in $\mathbf{H}$ were not required by symmetry to be degenerate and did not even have the same transformation properties under the operations of $\mathbf{H}$, that is, they belong to different irreducible representations, Γ_i and Γ_j, of $\mathbf{H}$. Once again $\langle u'_p{}^i |$ can be generated from $\langle u_p{}^i |$ by using eqn. (4.43).

In case (*b*) the wave functions for the term $D\Gamma_i$ and in case (*c*) the wave functions for the term $D\Gamma_{i,j}$ are the components of the row vector $\langle u_p{}^i, u'_p{}^i |$, where $\langle u_p{}^i |$ is assumed known from **H**. $\langle u'_p{}^i |$ could also be determined from first principles using the fact that it must form a basis for the representation $\bar{\Gamma}_i$ of the unitary subgroup **H**. However, when we have two terms sticking together in a case (*b*) or case (*c*) co-representation of **M** where the two terms were non-degenerate in **H**, there may be several different terms belonging to each of the representations Γ_i and $\bar{\Gamma}_i$ of **H**. In order to ensure that the correct pairs are taken together it may be safer to use eqn. (4.43) to generate $\langle u'_p{}^i |$ from $\langle u_p{}^i |$ rather than to determine $\langle u'_p{}^i |$ as a basis of $\bar{\Gamma}_i$ separately *ab initio*.

As an example of the determination of symmetry-adapted functions for the co-representations of a magnetic point group we consider the example of $4'/mmm'$ which has mmm as its unitary subgroup, **H**. The basis functions for the single-valued representations of **H** can be obtained from table 11 and the fact that the $+$ and $-$ signs on the labels of the representations of mmm go with even and odd values of *l*, respectively. We have already indicated in table 17 which of the representations of mmm lead to case (*a*) co-representations of $4'/mmm'$ and which lead to case (*c*) co-representations of $4'/mmm'$. The functions that belong to $\Gamma_1{}^+$, $\Gamma_1{}^-$, $\Gamma_3{}^+$ and $\Gamma_3{}^-$ therefore form bases of $D\Gamma_1{}^+$, $D\Gamma_1{}^-$, $D\Gamma_3{}^+$ and $D\Gamma_3{}^-$, respectively. The functions that belong to $\Gamma_2{}^+$ and $\Gamma_4{}^+$ or $\Gamma_2{}^-$ and $\Gamma_4{}^-$ stick together in pairs to form bases $\langle u_p{}^i, u'_p{}^i |$ of $D\Gamma_{2,4}{}^+$ or $D\Gamma_{2,4}{}^-$, respectively. It would seem reasonable to expect that the pairing of the functions will occur in such a manner that the two functions in any given pair will have common values of *l* and *m* ; thus, for example, we would expect $Y_1{}^{1,\,c}(\theta, \phi)$ and $Y_1{}^{1,\,s}(\theta, \phi)$ to come together to form one basis of $D\Gamma_{2,4}{}^-$, while $Y_2{}^{1,\,c}(\theta, \phi)$ and $Y_2{}^{1,\,s}(\theta, \phi)$ would come together to form a second basis of $D\Gamma_{2,4}{}^+$. This could be demonstrated directly, for example, by using θC_{2a} as A in eqn. (4.43) and taking $\langle u_p{}^i |$ to be one of the functions given in the table as belonging to $\Gamma_2{}^+$ or $\Gamma_2{}^-$ of **H** ; $A\langle u_p{}^i |$ can then be evaluated by using the known transformation properties of the spherical harmonics under operations of the rotation group $O(3)$ since C_{2a} is an element of $O(3)$ (see, for example, Cracknell (1966 a)).

§ 5. SPACE-GROUP SYMMETRY

5.1. *The determination and labelling of space-group irreducible representations*

We have already mentioned in § 1.4 that character tables for the irreducible representations of three important symmorphic space groups $Pm3m$ ($O_h{}^1$), $Fm3m$ ($O_h{}^5$) and $Im3m$ ($O_h{}^9$) were published by Bouckaert *et al.* (1936) (see also § 2.2). Character tables for two important non-symmorphic space groups $P6_3/mmc$ ($D_{6h}{}^4$), the space group of the h.c.p. structure, and $Fd3m$ ($O_h{}^7$), the space group of the diamond structure, were published by Herring (1942). These two publications were the first giving tables for irreducible representations for symmorphic and non-symmorphic space groups, respectively. Over the years the irreducible representations of the three-dimensional space groups have been used extensively to provide a scheme for labelling the electronic band structures of many important crystalline solids (see § 2.2). Moreover, it is also possible to use these irreducible representations to identify any essential degeneracies which must exist in a band structure and also, by generating symmetry-adapted functions, to simplify the calculation of band structures. Although it was originally concerned mainly with electronic band structures, the general theory of space-group representations is relevant to the study of the energies, as functions of $\mathbf{k}$, for any other particle, or quasi-particle excitation, in a crystalline solid. We shall therefore devote some attention to the problem of the determination and labelling of the space-group representations. Over the years there have been many papers published giving character tables for various selections of space groups. Recently, there have been some more systematic studies of space-group representations and, during the last decade, several complete sets of tables of irreducible representations of all the 230 space groups have been published (Faddeyev 1964, Kovalev 1965 a, Miller and Love 1967, Zak *et al.* 1969, Bradley and Cracknell 1972).

Although the principles involved in the determination of the irreducible representations of the space groups have been appreciated for over 30 years and although several sets of tables of these representations for all the 230 space groups have been published during the last decade, there is still no generally accepted scheme in use for the labelling of these representations. The first cause of the difficulty lies in the lack of uniqueness in the choice of the Brillouin zone and the labelling of the wave vectors $\mathbf{k}$, see § 1.4. A second difficulty arises even in those cases in which there is almost universal agreement about the shape of the Brillouin zone and the labels of the special points of symmetry ; this concerns the labelling of the irreducible representations for any given $\mathbf{k}$. Since the irreducible representations of a symmorphic space group are related in a rather trivial manner to the irreducible representations of the 32 point groups, see, for example, table 4 for $Fm3m$, it would seem to be natural to take over the labels used for the point-group representations to label the space-group representations as well. However, this presents difficulties because there is more than one quite well-established scheme in use for labelling point-group irreducible representations. These include the Mulliken (1933) notation which is very widely used, particularly among chemists ; this scheme has the advantage that many useful properties of a representation, such as its dimension, parity, and whether it is single-valued or double-valued, are immediately indicated by the label used. Other schemes, such as that of Bouckaert *et al.* (1936), are based on the use of labels Γ_i, with the symbol Γ replaced by the label of the special wave vector in question ; the subscript i discriminates among the irreducible representations of a given point group. Various forms of the Γ_i notation exist but the most standard form is probably that used by Koster *et al.* (1963). In addition to this, the orientations in which the point groups may arise in connection with representations of symmorphic space groups may differ from the orientations of those point groups in the standard tables of point-group representations.

In dealing with non-symmorphic space groups it is common to label the irreducible representations numerically (see, for example, Herring (1942)) ; thus, for example, at the point X in the Brillouin zone in fig. 8 the irreducible representations would be labelled as X_1, X_2, X_3, ... etc. However, there is little agreement among the various sets of tables about the sequence used in numbering the representations for any given $\mathbf{k}$ (see the list of tables of space-group representations which we have already mentioned). Attempts have also been made to extend the Mulliken notation to the non-symmorphic space groups (see, for example, Altmann and Bradley (1965), Bradley and Cracknell (1972)); however, such labels tend to become rather cumbersome and it is also difficult to be completely systematic in the assignment of these labels. In view of these difficulties and also of the fact that a simple numerical labelling scheme is better for use in computer generation and storage of irreducible representations, it seems unlikely that the extended Mulliken scheme will ever be widely adopted. A further difficulty which arises when one tries to extend the Mulliken notation to the labelling of space-group representations, or to develop any other systematic set of rules for labelling these representations, is that the behaviour of these representations for symmetry operations $\{R|\mathbf{v}\}$ may vary according to one's choice of the co-ordinate system which, of course, affects the values of many of the vectors $\mathbf{v}$ (see, for example, Cornwell (1971, 1972)). Thus, suppose that with one choice of origin we have identified all the Seitz symbols $\{R|\mathbf{v}\}$ for a given space group, **G**, and have determined and labelled all the irreducible repre-sentations of **G**. If we now move the origin by $\{E|\mathbf{t}_0\}$, where $\mathbf{t}_0$ is not necessarily a translation of the Bravais lattice, the forms of the Seitz symbols will be altered to $\{R|\mathbf{v}'\}$ where

$$\{R|\mathbf{v}'\} = \{R|\mathbf{v} + R\mathbf{t}_0 - \mathbf{t}_0\} \tag{5.1}$$

(see eqn. (3.5.10) of Bradley and Cracknell (1972)). Thus the matrix repre-sentative, or the character, of $\{R|\mathbf{v}\}$ in $\Gamma_p^{\ \mathbf{k}}$ with the old choice of origin becomes the matrix representative, or the character, of $\{R|\mathbf{v} + R\mathbf{t}_0 - \mathbf{t}_0\}$ in $\Gamma_p^{\ \mathbf{k}}$ with the new choice of origin. There is no unique standard choice of origin for each of the space groups and even the International Tables (Henry and Lonsdale 1965) give two different choices of origin for many of the space groups. It is also possible to change the forms of the Seitz space-group symbols by altering the choice of the orientation of the x, y and z axes rela-tive to the axes and planes of symmetry in a crystal. Consequently, with a given set of rules for labelling the irreducible representations, based on the characters of $\{R|\mathbf{v}\}$ or $\{R|\mathbf{v}'\}$ for certain important choices of R, one may find some permutation of the set of labels for any given $\mathbf{k}$ if the origin is moved and the rules are applied a second time. In view of this and all the other difficulties mentioned previously, it may well be that a simple number-ing scheme, such as that used by Miller and Love (1967), in which one does not attempt even to indicate the parity of a representation, might be the easiest to adopt and use if it could be accepted as the standard scheme for labelling space-group representations. However it will be difficult to displace existing established sets of labels for the irreducible representations of some of the more common space groups such as $Fm3m$ $(O_h^{\ 5})$ which was included as an example in table 4.

The question of the standardization of labels for the irreducible co-representations of the Heesch–Shubnikov space groups presents less difficulty than is the case for the irreducible representations of the classical groups. There is only one published set of tables which gives the irreducible co-representations for all the non-unitary Heesch–Shubnikov space groups (Miller and Love 1967), although other sets of tables do exist for a few particular groups. This scheme is based on a simple numbering for all the irreducible co-representations for each wave vector $\mathbf{k}$. This is convenient for the computer generation and storage of the tables but has the disadvantage that the labels convey little information about the co-representations themselves, such as for instance indicating whether a given co-representation is single-valued or double-valued. In particular this scheme for labelling the irreducible co-representations has the disadvantage that it does not indicate, for any given co-representations, which are the irreducible representations of $\mathbf{H}$, the unitary sub-group, from which those co-representations are derived.

5.2. *Time-reversal symmetry and space-group representations*

In the various tables of space-group representations which exist (Faddeyev 1964, Kovalev 1965 a, Miller and Love 1967, Zak *et al.* 1969, Bradley and Cracknell 1972) there are tabulated the representations of $\mathbf{G^k}$. The degeneracies of the energy levels $E_j(\mathbf{k})$ of particles or quasiparticles with wave vector $\mathbf{k}$ are then immediately available ; they are just the degeneracies of $\Gamma_j{}^{\mathbf{k}}$. However, if we are considering a crystal in which there is no magnetic ordering and which is not situated in an external magnetic field, the crystal will actually possess time-reversal symmetry in addition to all the spatial symmetry operations of $\mathbf{G}$. The complete group of the symmetry operations of the crystal is then a grey space group, which is the direct product $\mathbf{G} \otimes (E + \theta)$, where θ is the operation of time inversion. The addition of the extra symmetry operation θ may cause some extra degeneracies in the energy levels of a particle or quasiparticle with wave vector $\mathbf{k}$ (Herring 1937, Frei 1966, Bradley and Cracknell 1970, 1972, Backhouse 1973). If $\Gamma_j{}^{\mathbf{k}}$ is an irreducible representation of $\mathbf{G^k}$ then there are several possibilities :

(*a*) there is no change in the degeneracy of $\Gamma_j{}^{\mathbf{k}}$;

(*b*) the degeneracy of $\Gamma_j{}^{\mathbf{k}}$ becomes doubled, or in terms of energy levels, two different energy levels, both described by $\Gamma_j{}^{\mathbf{k}}$, become degenerate ;

(*c*) the degeneracy of $\Gamma_j{}^{\mathbf{k}}$ becomes doubled but, unlike (*b*), two different, that is inequivalent, irreducible representations $\Gamma_j{}^{\mathbf{k}}$ and $\Gamma_{j'}{}^{\mathbf{k}}$ of $\mathbf{G^k}$ become degenerate ;

these effects arise because of the addition of θ to the space group of the crystal (see § 1.2).

In addition to the possibility of causing some extra degeneracies in the spectrum of the energy eigenvalues at $\mathbf{k}$, the inclusion of time-reversal symmetry will always cause degeneracies between $\mathbf{k}$ and $-\mathbf{k}$ when these two vectors do not appear in the same star (for details of proof see, for example, Bradley and Cracknell (1970, 1972)). Of course, if $\mathbf{k}$ and $-\mathbf{k}$ do appear in the same star, as for example when the crystal has a centre of inversion symmetry so that $\{I|\mathbf{v}\}$ is in $\mathbf{G}$, the spectrum of the energy eigenvalues at $\mathbf{k}$ is the same as at $-\mathbf{k}$ even in the absence of time-reversal symmetry. This extra degeneracy, which we shall call a type (*x*) degeneracy, is very easy to recognize as it occurs when time-reversal symmetry is present and there are no space-group elements that transform $\mathbf{k}$ into $-\mathbf{k}$.

The most convenient way of considering these extra degeneracies is by considering the theory of the co-representations of the grey space group $(\mathbf{G} + \theta\mathbf{G})$. It should be noted, of course, that strictly speaking it is the induced representations $(\Gamma_j{}^\mathbf{k}\!\uparrow\!\mathbf{G})$ and the corresponding induced co-representations in $(\mathbf{G}+\theta\mathbf{G})$ that we are really concerned with. For the grey space group $\mathbf{M}=\mathbf{G}+\theta\mathbf{G}$ the magnetic little group $\mathbf{M}^\mathbf{k}$, for a given wave vector $\mathbf{k}$, takes one of three forms :

$$\mathbf{M}^\mathbf{k}=\mathbf{G}^\mathbf{k} \tag{5.2}$$

$$\mathbf{M}^\mathbf{k}=\mathbf{G}^\mathbf{k}+\theta\mathbf{G}^\mathbf{k} \tag{5.3}$$

and

$$\mathbf{M}^\mathbf{k}=\mathbf{G}^\mathbf{k}+A\,\mathbf{G}^\mathbf{k} \tag{5.4}$$

where A is some anti-unitary element which is a product of θ and some space-group element other than a translation (that is A cannot be chosen to be $\theta\{E\,|\,\boldsymbol{\tau}\}$). The unitary elements of $\mathbf{M}^\mathbf{k}$ transform $\mathbf{k}$ into $+\mathbf{k}$ while the space-group parts of the anti-unitary elements of $\mathbf{M}^\mathbf{k}$ transform $\mathbf{k}$ into $-\mathbf{k}$. When we referred to the three possible situations (a), (b) or (c) which might arise and affect the degeneracy of $\Gamma_j{}^\mathbf{k}$, we were actually using the notation of case (a), case (b) and case (c) co-representations of the grey group $(\mathbf{G}+\theta\mathbf{G})$ (see § 1.3).

If $\mathbf{M}^\mathbf{k}$ is given by eqn. (5.2) there are no elements in $\mathbf{G}$ which transform $\mathbf{k}$ into $-\mathbf{k}$. This only happens for $\mathbf{k}$ vectors for which $-\mathbf{k}$ does not appear in the star of $\mathbf{k}$. Clearly $\mathbf{M}^\mathbf{k}$ contains no anti-unitary operators, and the degeneracies of the eigenvalues in $\mathbf{M}^\mathbf{k}$ are the same as in $\mathbf{G}^\mathbf{k}$, but there is a degeneracy between each of the eigenvalues at $\mathbf{k}$ and one of the eigenvalues at $-\mathbf{k}$. That is, the spectrum of the eigenvalues at $\mathbf{k}$ is the same as that at $-\mathbf{k}$ as a result of time-reversal symmetry, even though $-\mathbf{k}$ is not in the star of $\mathbf{k}$; in other words there is a type (x) degeneracy at $\mathbf{k}$. Such situations are identified explicitly as type (x) degeneracies in the tables of Bradley and Cracknell (1972) and they are also very easy to recognize in the tables of Miller and Love (1967), where under the entry in table V for the appropriate grey space group one finds the entry ' GROUP K IS UNITARY '.

If $\mathbf{M}^\mathbf{k}$ takes the form in eqn. (5.3) or eqn. (5.4) the irreducible co-representation of $\mathbf{M}$ derived from the induced representation $(\Gamma_j{}^\mathbf{k}\!\uparrow\!\mathbf{G})$ of $\mathbf{G}$ may belong to case (a), case (b) or case (c). For any given representation $\Gamma_j{}^\mathbf{k}$ the case which actually applies can be determined by using a well-known test, see, for example, eqn. (2.40) of Bradley and Davies (1968) or eqn. (7.6.16) of Bradley and Cracknell (1972) ; this has been done and the results were included in the various published sets of tables of space-group representations. For both eqn. (5.3) and eqn. (5.4) $-\mathbf{k}$ is in the star of $\mathbf{k}$ so that the spectra of eigenvalues at $\mathbf{k}$ and $-\mathbf{k}$ are already identical when only the symmetry of $\mathbf{G}$ is considered. The only additional degeneracy that can arise because of the presence of the additional elements $\theta\mathbf{G}$ in $\mathbf{M}$ will involve extra degeneracies among the various eigenvalues at $\mathbf{k}$. If the symmetry of a crystal is described by $\mathbf{G}$, the degeneracies of the eigenvalues for particles or quasi-particles with wave vector $\mathbf{k}$ will be determined by the irreducible representations $\Gamma_j{}^\mathbf{k}$ of $\mathbf{G}^\mathbf{k}$; if the symmetry of the crystal is described by the grey group $\mathbf{M}=\mathbf{G}+\theta\mathbf{G}$ the degeneracies of the eigenvalues will be determined by the co-representations $D\Gamma_j{}^\mathbf{k}$ of $\mathbf{M}^\mathbf{k}$. In case (a) there is no increase in degeneracy at $\mathbf{k}$, but in cases (b) and (c) a doubling of the degeneracy occurs. In case (b), $D\Gamma_j{}^\mathbf{k}$ contains $\Gamma_j{}^\mathbf{k}$ twice and this implies that at $\mathbf{k}$ the extra degeneracy is between two sets of eigenvalues both belonging to the same representation

$\Gamma_j{}^{\mathbf{k}}$. In case (c), $D\Gamma_j{}^{\mathbf{k}}$ contains $\Gamma_j{}^{\mathbf{k}}$ and $\overline{\Gamma}_j{}^{\mathbf{k}}$; $\overline{\Gamma}_j{}^{\mathbf{k}}$ is not equivalent to $\Gamma_j{}^{\mathbf{k}}$ and must, therefore, be equivalent to some other representation $\Gamma_{j'}{}^{\mathbf{k}}$ of $\mathbf{G}^{\mathbf{k}}$ $(j' \neq j)$. Thus, in case (c) the extra degeneracy is between two sets of eigenvalues belonging to different representations of $\mathbf{G}^{\mathbf{k}}$.

5.3. *Irreducible representations of the two-dimensional space groups—surfaces and thin films*

In view of the very large amount of work which has been done on the representations of the three-dimensional space groups and which we have already discussed, it is perhaps surprising that relatively little work has been devoted to the determination of the irreducible representations of the two-dimensional space groups. The reason for this is presumably that in the early days of the study of the theory of solid-state physics people were concerned with particles, or quasiparticle excitations, in specimens of solids that were of large extent in each of three non-coplanar directions. There seemed to be no immediate use for the irreducible representations of the two-dimensional space groups. However, recently there has been a substantial rise of interest in surface states in crystalline solids and also in the behaviour of particles or quasiparticles in thin films of crystalline materials. If a particle or quasiparticle is constrained to move so that it remains on the surface of a specimen of a crystalline solid, or if the specimen is in the form of a thin film, it will become necessary to use a two-dimensional space group instead of a three-dimensional space group. By 'thin' in this context we mean that the thickness of the film is equal to only a small number of unit cell repeat distances. There is a slight difference between a surface state and a thin-film state. In a surface state the wave function is an evanescent wave in the direction normal to the surface, that is the component of $\mathbf{k}$ normal to the surface is pure imaginary. In the case of a thin film, where the thickness of the film is only of the order of a few unit cell lengths, the wave functions form a standing-wave pattern across the film so that the values of the allowed component of $\mathbf{k}$ across the film are determined geometrically by the boundary conditions and the thickness of the film. However, for the plane in which there is the two-dimensional translational symmetry, the surface and thin-film wave functions both have the same Bloch-wave form. In the following discussion we shall frequently use the word 'surface' to include both a surface and a thin-film situation.

We give a list of the 17 two-dimensional space groups in table 19. In considering surface states we assume that the surface in question is a possible natural face of the crystal under investigation. That is, it can be described or labelled by a set of Miller indices (hkl) or $(hkil)$, where h, k, i and l are rather small integers. Similarly, in considering crystalline thin films we assume that the plane of a thin film is parallel to a possible natural face of the crystalline material. If we do not satisfy these conditions for surface states and thin-film states, respectively, the symmetry of the surface or thin film will not be described by one of the two-dimensional space groups. Of course, any two-dimensional space group, $\mathbf{G}^s$, which describes the symmetry of a surface (hkl) of a given crystal will be a subgroup of the three-dimensional space group, $\mathbf{G}$, which describes the symmetry of a bulk specimen of that

Table 19.　The 17 two-dimensional space groups

System	Number	Short symbol	Full symbol	Generating elements	S/NS
Oblique, p	1	$p1$	$p1$	$\{E\|\mathbf{0}\}$	S
	2	$p2$	$p211$	$\{C_{2z}\|\mathbf{0}\}$	S
Rectangular, p or c	3	pm	$p1m1$	$\{\sigma_y\|\mathbf{0}\}$	S
	4	pg	$p1g1$	$\{\sigma_y\|\frac{1}{2}\mathbf{t}_1{}^s\}$	NS
	5	cm	$c1m1$	$\{\sigma_x\|\mathbf{0}\}$	S
	6	pmm	$p2mm$	$\{C_{2z}\|\mathbf{0}\}, \{\sigma_y\|\mathbf{0}\}$	S
	7	pmg	$p2mg$	$\{C_{2z}\|\mathbf{0}\}, \{\sigma_y\|\frac{1}{2}\mathbf{t}_1{}^s\}$	NS
	8	pgg	$p2gg$	$\{C_{2z}\|\mathbf{0}\}, \{\sigma_x\|\frac{1}{2}\mathbf{t}_1{}^s+\frac{1}{2}\mathbf{t}_2{}^s\}$	NS
	9	cmm	$c2mm$	$\{C_{2z}\|\mathbf{0}\}, \{\sigma_x\|\mathbf{0}\}$	S
Square, p	10	$p4$	$p4$	$\{C_{4z}{}^+\|\mathbf{0}\}$	S
	11	$p4m$	$p4mm$	$\{C_{4z}{}^+\|\mathbf{0}\}, \{\sigma_x\|\mathbf{0}\}$	S
	12	$p4g$	$p4gm$	$\{C_{4z}{}^+\|\mathbf{0}\}, \{\sigma_x\|\frac{1}{2}\mathbf{t}_1{}^s+\frac{1}{2}\mathbf{t}_2{}^s\}$	NS
Hexagonal, p	13	$p3$	$p3$	$\{C_3{}^+\|\mathbf{0}\}$	S
	14	$p3m1$	$p3m1$	$\{C_3{}^+\|\mathbf{0}\}, \{\sigma_{v1}\|\mathbf{0}\}$	S
	15	$p31m$	$p31m$	$\{C_3{}^+\|\mathbf{0}\}, \{\sigma_{d1}\|\mathbf{0}\}$	S
	16	$p6$	$p6$	$\{C_6{}^+\|\mathbf{0}\}$	S
	17	$p6m$	$p6mm$	$\{C_6{}^+\|\mathbf{0}\}, \{\sigma_{d1}\|\mathbf{0}\}$	S

Notes. (i). The space groups are labelled in columns 2, 3 and 4 according to the international tables (Henry and Lonsdale 1965). (ii). In the labelling of the generating elements of pm and pg we have departed from the orientation used in the International Tables ; the reason for this is to preserve pg as a subgroup of pmg. In this orientation the full symbols should perhaps be $p11m$ and $p11g$, respectively. (iii). The symbol NS or S in the last column indicates whether the appropriate space group is non-symmorphic or symmorphic, respectively.

crystal. It would therefore, in principle, be possible to determine the irreducible representations of any given $\mathbf{G}^s$ by exploiting this subgroup relation between $\mathbf{G}$ and $\mathbf{G}^s$. However, this may often be quite difficult and it is easier to determine the irreducible representations of the two-dimensional space groups $\mathbf{G}^s$ for surface states or thin-film states directly from first principles (Jones 1966, Cracknell 1974 b).

In the conventional theory to which we have alluded in § 5.1 it is assumed that one is concerned with a specimen which is, from the microscopic point of view, of very large extent in all directions. If the specimen is assumed to be of infinite extent, then its symmetry can be described by one of the 230 three-dimensional space groups. However, in our present situation we are concerned with Hamiltonians with periodicity in only two dimensions instead of three dimensions. In the case of surface states the Hamiltonian has translational symmetry in two dimensions in the plane of the surface but has no translational symmetry in a direction out of the plane of the surface. In the case of a thin film the Hamiltonian has translational symmetry in two dimensions in the plane of the film but does not possess translational symmetry across the thickness of the film. That is, we have a material which one might regard as a crystalline solid in two directions and as a molecule in the third direction. If $\mathbf{t}_1{}^s$ and $\mathbf{t}_2{}^s$ are the basic vectors of the two-dimensional translational symmetry of the Hamiltonian of a surface eigenstate or a thin-film eigenstate, the eigenfunctions will have Bloch-type behaviour in the plane of the surface or thin film but not in any direction that is not in this plane. By analogy with the three-dimensional case the wave vector $\mathbf{k}^s$ in this plane takes the form

$$\mathbf{k}^s = k_1{}^s\mathbf{g}_1{}^s + k_2{}^s\mathbf{g}_2{}^s \tag{5.5}$$

where

$$\mathbf{g}_i{}^s \cdot \mathbf{t}_j{}^s = 2\pi\delta_{ij} \tag{5.6}$$

and

$$k_i{}^s = p_i{}^s/N_i{}^s \tag{5.7}$$

($i = 1, 2$). We have used the superscript s on the symbols $\mathbf{t}_i{}^s$, $\mathbf{g}_i{}^s$, $k_i{}^s$, $p_i{}^s$ and $N_i{}^s$ in these equations for surface or thin-film states to emphasize that these symbols are not necessarily identical with the corresponding symbols used in eqns. (1.14)–(1.17) for the three-dimensional case. The precise relationship between the vectors $\mathbf{t}_i{}^s$ for a surface and $\mathbf{t}_i$ for a bulk specimen of the same material will depend on the form (in the crystallographic sense) of the face which constitutes the surface.

All the physically distinguishable states of a system with two-dimensional translational symmetry can be described in relation to the appropriate two-dimensional Brillouin zone, which is just a unit cell of the vector space based on $\mathbf{g}_1{}^s$ and $\mathbf{g}_2{}^s$. The two-dimensional Brillouin zones are illustrated in fig. 13.

The translational symmetry operations of the two-dimensional space group of a surface will be those translational operations of $\mathbf{G}$ that are in the plane of the surface. In addition to the translational symmetry operations there are also the possibilities of

(i) the identity operation,

(ii) pure rotations about an axis normal to the surface,

(iii) reflections in planes normal to the surface,

(iv) glide reflections in planes normal to the surface, where the glide vector is parallel to the surface.

The presence of these symmetry operations can be detected by inspection of the entry in the international tables (Henry and Lonsdale 1965) for $\mathbf{G}$, albeit perhaps with a little difficulty for some of the more complicated groups, especially the cubic groups.

Let us consider in a little more detail the relationship between the two-dimensional space groups $\mathbf{G}^s$ for the surfaces of a given crystal and the three-dimensional space group $\mathbf{G}$ of that crystal. An example may be helpful in illustrating what is involved. Suppose we consider surface states of a crystal with the cubic face-centred (F) Bravais lattice ; such crystals include those with the sodium chloride structure or the diamond structure. For (001), (110) and (111) faces we may choose $\mathbf{t}_1{}^s$ and $\mathbf{t}_2{}^s$ and then use eqn. (5.6) to obtain $\mathbf{g}_1{}^s$ and $\mathbf{g}_2{}^s$:

$$
\begin{array}{cccc}
 & (001) & (110) & (111) \\[4pt]
\mathbf{t}_1{}^s & a(\tfrac{1}{2}, \tfrac{1}{2}, 0) & a(-\tfrac{1}{2}, \tfrac{1}{2}, 0) & a(\tfrac{1}{2}, -\tfrac{1}{2}, 0) \\[6pt]
\mathbf{t}_2{}^s & a(\tfrac{1}{2}, -\tfrac{1}{2}, 0) & a(0, 0, 1) & a(\tfrac{1}{2}, 0, -\tfrac{1}{2}) \\[6pt]
\mathbf{g}_1{}^s & \dfrac{2\pi}{a}(1, 1, 0) & \dfrac{2\pi}{a}(1, 1, 0) & \dfrac{2\pi}{a}(\tfrac{2}{3}, -\tfrac{4}{3}, \tfrac{2}{3}) \\[10pt]
\mathbf{g}_2{}^s & \dfrac{2\pi}{a}(1, -1, 0) & \dfrac{2\pi}{a}(0, 0, 1) & \dfrac{2\pi}{a}(\tfrac{2}{3}, \tfrac{2}{3}, -\tfrac{4}{3}).
\end{array}
$$

Fig. 13

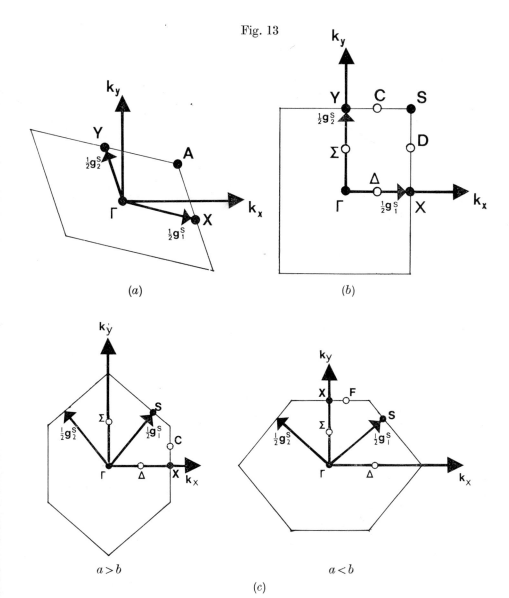

(a) (b)

$a > b$ $a < b$

(c)

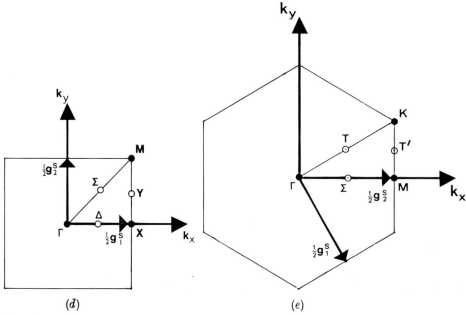

(d)   (e)

Two-dimensional Brillouin zones : (a) oblique, p, (b) rectangular, p, (c) rectangular, c, (d) square, p, (e) hexagonal, p.

Recalling that the basic vectors of the reciprocal lattice for the cubic F lattice may be taken as

$$\mathbf{g}_1 = \frac{2\pi}{a}(-1, 1, 1)$$

$$\mathbf{g}_2 = \frac{2\pi}{a}(1, -1, 1) \qquad (5.8)$$

$$\mathbf{g}_3 = \frac{2\pi}{a}(1, 1, -1)$$

we see that in each case $\mathbf{g}_1{}^s$ and $\mathbf{g}_2{}^s$ are not translation vectors of the three-dimensional reciprocal lattice. (Certain multiples or combinations of $\mathbf{g}_1{}^s$ and $\mathbf{g}_2{}^s$ will, however, be translation vectors of the three-dimensional reciprocal lattice, namely $p\mathbf{g}_1{}^s + q\mathbf{g}_2{}^s$, where p and q are integers and $(p+q)$ is a multiple of 2 for the (001) and (110) surfaces and $(p+q)$ is a multiple of 3 for the (111) surface.) Thus, although we can regard the two-dimensional Brillouin zone for (hkl) surface states in a cubic F crystal as being in the cross section of the three-dimensional Brillouin zone parallel to (hkl) and passing through $\mathbf{k}=\mathbf{0}$, the boundaries of this two-dimensional Brillouin zone need not coincide with the boundaries of the three-dimensional Brillouin zone. This is illustrated in fig. 14, where the boundaries have been identified by drawing the perpendicular bisectors of the two-dimensional reciprocal lattice vectors. One or two interesting features may be noticed. Wave vectors which were not equivalent in the three-dimensional Brillouin zone may become equivalent in the two-dimensional case. For the (111) surface states the points at the corners

Fig. 14

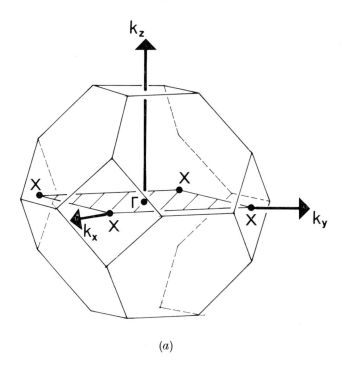

(a)

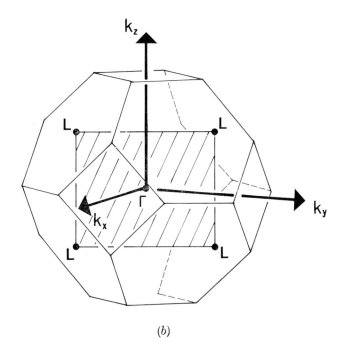

(b)

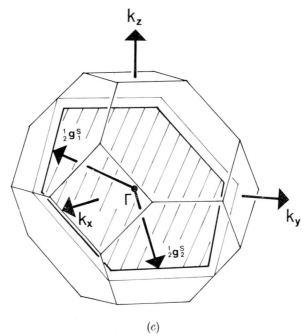

(c)

Two-dimensional Brillouin zones (shaded) inscribed in the three-dimensional Brillouin zone for (a) (001), (b) (110) and (c) (111) surfaces on cubic F crystals.

of the two-dimensional Brillouin zone, which are K points in the notation of fig. 13 (e), have more symmetry than the same wave vector would have in the three-dimensional Brillouin zone. The cause of these peculiarities can be traced back to the fact that $g_1{}^s$ and $g_2{}^s$ are not translation vectors of the three-dimensional reciprocal lattice. Thus for a (111) surface it would be rather difficult to try to determine the irreducible representations of $\mathbf{G}^s$ from the tabulated irreducible representations of $\mathbf{G}$. In practice when considering (100), (110) and (111) surface states in silicon, which has the diamond structure, Jones (1966) found it much more convenient to determine the irreducible representations of $\mathbf{G}^s$ *ab initio* rather than to try to determine them from the tables for the appropriate three-dimensional space group $Fd3m$.

The determination of the irreducible representations of the two-dimensional space groups can be performed in a manner which is directly analogous to that used in the determination of the irreducible representations of the three-dimensional space groups. For a symmorphic two-dimensional space group the irreducible representations of the space group corresponding to $\mathbf{k}^s$ will be related in a trivial manner to the irreducible representations of one of the classical point groups which we may call $\mathbf{P}(\mathbf{k}^s)$. Since $\mathbf{P}(\mathbf{k}^s)$ is a classical point group the problem of the determination of the irreducible representations of the symmorphic space groups reduces to looking up the irreducible representations of the appropriate point group in some standard set of tables of point-group representations. In such tables the representations may be labelled as $\Gamma_1, \Gamma_2, ..., \Gamma_j, ...$ (see, for example, Koster *et al.* (1963)). However, when considered as representations of $\mathbf{P}(\mathbf{k}^s)$ the symbol Γ is replaced by the symbol used to label the special point or line of symmetry corresponding to

$\mathbf{k}^s$. The details of the determination of the irreducible representations of the four two-dimensional space groups *pg*, *pmg*, *pgg* and *p4g*, which are non-symmorphic, are considered in a paper by Cracknell (1974 b) which contains complete tables for these groups ; the extension of this work to the two-dimensional Heesch–Shubnikov groups, for materials that exhibit magnetic ordering, has also been considered (Cracknell 1974 d).

5.4. *Space-group symmetry for a crystal in an applied magnetic field*

The theory of space groups and of their irreducible representations, or co-representations, which we have used so far has been concerned with crystals which are not situated in an external electric or magnetic field. The Hamiltonian, $\mathscr{H}$, of an electron, or of some quasiparticle, in such a crystal can be assumed to possess the full symmetry of the space group of the crystal, because the potential $V(\mathbf{r})$ has the full symmetry of the crystal.

Let us now consider the effect of an external electric or magnetic field on the quantum-mechanical eigenstates of the Hamiltonian of a particle or quasiparticle in a crystal. If the external field is non-uniform the space-group symmetry of the crystal will generally be incompatible with any symmetry that may be possessed by the external field. The potential term $V(\mathbf{r})$ in the Hamiltonian will vary from one lattice point to another and therefore the Hamiltonian will no longer possess the symmetry of the space group **G** of the crystal. This loss of the translational invariance of $\mathscr{H}$ invalidates the whole edifice of the use of **k** as a good quantum number and of the irreducible representations of **G** as the basis of a scheme for classifying and labelling the eigenstates of $\mathscr{H}$. If the external field is uniform the potential $V(\mathbf{r})$ will still vary from one lattice point to another so that $\mathscr{H}$ will not possess the translational invariance of **G**, because it is the potential and not the field which appears in $\mathscr{H}$. Therefore, even for a uniform field, the use of **k** and of the conventional irreducible representations of **G** will not be expected to apply. However, the fact that a uniform field possesses the translational invariance of **G**, whereas a non-uniform field does not, means that there are some symmetry arguments which can be used in the uniform-field case but which will not apply for a non-uniform field (Wannier and Fredkin 1962). Thus it has been suggested that for an electron in a crystal that is placed in an external field, the eigenfunctions of the Hamiltonian will be similar to Bloch eigenfunctions and will reduce to Bloch eigenfunctions as the magnitude of the external field is reduced to zero. It has been found that the eigenfunctions of $\mathscr{H}$ belong to certain representations which are called ' projective representations ' of the space group, **G**, of the crystal. In this section we shall indicate some of the ideas involved in the construction of projective representations, following particularly the work of Brown (1964), Zak (1964 a, b), Ashby and Miller (1965), Rudra (1965) ; for a more extensive discussion of the formal theory we would recommend the series of papers by Backhouse and Bradley (Backhouse 1970, 1971, Backhouse and Bradley 1970, 1972, Bradley 1973).

Let us consider an electron in a periodic potential in a uniform magnetic field. Suppose that the Hamiltonian

$$\mathscr{H} = p^2/2m + V(\mathbf{r}) \tag{5.9}$$

describes the motion of an electron in a periodic potential $V(\mathbf{r})$ in the absence of an external magnetic field. Then the corresponding Hamiltonian when the external uniform magnetic field is included can be obtained by replacing $\mathbf{p}$ by $(\mathbf{p}+e\mathbf{A})$ so that

$$\mathscr{H} = (\mathbf{p}+e\mathbf{A})^2/2m + V(\mathbf{r}). \qquad (5.10)$$

$\mathbf{A}$ is the magnetic vector potential and can be taken to be $-\frac{1}{2}(\mathbf{r}\wedge\mathbf{B})$. In the absence of an external magnetic field the effect of a translation operation $\{E|\mathbf{T_n}\}$ $(=\{E|n_1\mathbf{t}_1+n_2\mathbf{t}_2+n_3\mathbf{t}_3\})$ on a Bloch function is to multiply the Bloch function by a factor $\exp[-i\mathbf{k}\,.\,\mathbf{T_n}]$, that is

$$\{E|\mathbf{T_n}\}\psi(\mathbf{r}) = \psi(\mathbf{r}-\mathbf{T_n}). \qquad (5.11)$$

We can rewrite $\exp[-i\mathbf{k}\,.\,\mathbf{T_n}]$ as $\exp[-i(\mathbf{p}\,.\,\mathbf{T_n})/\hbar]$ and this suggests that when an external magnetic field is present we can define a magnetic translation operator which has the effect of multiplying an eigenfunction $\psi(\mathbf{r})$ by the factor $\exp[-i\{(\mathbf{p}+e\mathbf{A})\,.\,\mathbf{T_n}\}/\hbar]$ (Brown 1964). We shall use the symbol $\{E|M\mathbf{T_n}\}$ to denote this operator, where

$$\begin{aligned}\{E|M\mathbf{T_n}\}\psi(\mathbf{r}) &= \exp[-i\{(\mathbf{p}+e\mathbf{A})\,.\,\mathbf{T_n}\}/\hbar]\psi(\mathbf{r}) \\ &= \exp[(ie/2\hbar)(\mathbf{r}\wedge\mathbf{B})\,.\,\mathbf{T_n}]\psi(\mathbf{r}-\mathbf{T_n}) \\ &= \exp[\tfrac{1}{2}i(\mathbf{T_n}\wedge\boldsymbol{\beta})\,.\,\mathbf{r}]\psi(\mathbf{r}-\mathbf{T_n}) \qquad (5.12)\end{aligned}$$

where $\boldsymbol{\beta}=e\mathbf{B}/\hbar$. It is worth noting some of the properties of the magnetic translation operators $\{E|M\mathbf{T_n}\}$; thus

(i) $\{E|M\mathbf{T_n}\}$ commutes with the Hamiltonian in eqn. (5.10),

(ii) the magnetic translations do not commute with one another, thus it can be shown that

$$\begin{aligned}\{E|M\mathbf{T_m}\}\{E|M\mathbf{T_n}\} &= \exp[-i(\mathbf{T_m}\wedge\mathbf{T_n})\,.\,\boldsymbol{\beta}] \\ &\quad \times\{E|M\mathbf{T_n}\}\{E|M\mathbf{T_m}\} \qquad (5.13)\end{aligned}$$

(iii) the product of an arbitrary pair of the magnetic translation operations is not necessarily one of the magnetic translations,

$$\begin{aligned}\{E|M\mathbf{T_m}\}\{E|M\mathbf{T_n}\} &= \exp[(-i/2)(\mathbf{T_m}\wedge\mathbf{T_n})\,.\,\boldsymbol{\beta}] \\ &\quad \times\{E|M(\mathbf{T_m}+\mathbf{T_n})\}. \qquad (5.14)\end{aligned}$$

Properties (ii) and (iii) contrast with the situation for ordinary lattice translations $\{E|\mathbf{T_n}\}$ in the absence of an external field; the translations $\{E|\mathbf{T_m}\}$ and $\{E|\mathbf{T_n}\}$ commute so that the translational group is Abelian, while the operations $\{E|\mathbf{T_n}\}$ satisfy the closure condition and form a group. (ii) expresses the fact that the $\{E|M\mathbf{T_n}\}$ do not commute, while (iii) expresses the fact that the operations $\{E|M\mathbf{T_n}\}$ do not satisfy the condition of closure and therefore do not form a group.

Although the magnetic translation operations $\{E|M\mathbf{T_n}\}$ do not form a group it is still instructive to consider the effect of $\{E|M\mathbf{T_n}\}$ on the eigenfunctions of $\mathscr{H}$. Suppose that $\psi_m(\mathbf{r})$ is an eigenfunction of $\mathscr{H}$ with m-fold degeneracy, and that

$$\{E|M\mathbf{T_n}\}\psi_m(\mathbf{r}) = \sum_{l=1}^{M} D_{lm}(\{E|M\mathbf{T_n}\})\psi_l(\mathbf{r}). \qquad (5.15)$$

From eqns. (5.13) and (5.14) it follows that

$$\mathbf{D}(\{E\,|\,M\mathbf{T_m}\})\mathbf{D}(\{E\,|\,M\mathbf{T_n}\}) = \exp\left[-\tfrac{1}{2}i(\mathbf{T_m} \wedge \mathbf{T_n}) \cdot \boldsymbol{\beta}\right]$$
$$\times \mathbf{D}(\{E\,|\,M(\mathbf{T_m} + \mathbf{T_n})\}). \quad (5.16)$$

Thus the matrices $\mathbf{D}(\{E\,|\,M\mathbf{T_n}\})$ do not form an ordinary representation of the translational subgroup of $\mathbf{G}$. This is to be compared with the relation

$$\mathbf{D}(\{E\,|\,\mathbf{T_m}\})\mathbf{D}(\{E\,|\,\mathbf{T_n}\}) = \mathbf{D}(\{E\,|\,\mathbf{T_m} + \mathbf{T_n}\}) \quad (5.17)$$

which would hold for an ordinary representation of the translational subgroup of $\mathbf{G}$. The matrices which are defined by eqn. (5.15) and which have the multiplication properties given in eqn. (5.16) form what is called a *projective representation* of the translational subgroup of $\mathbf{G}$. The details of the determination of these projective representations for the translational subgroup of $\mathbf{G}$ have been considered by a number of authors (Brown 1964, Zak 1964 a, b, Backhouse and Bradley 1972).

The conventional way to construct the irreducible representations of the translational subgroup $\mathbf{T}$ of a space group $\mathbf{G}$ involves considering a finite piece of crystal and applying Born–von Kármán cyclic boundary conditions. For the case of a crystal in a magnetic field we therefore restrict our attention to a finite lattice of dimensions $N_1\mathbf{t_1}$, $N_2\mathbf{t_2}$ and $N_3\mathbf{t_3}$ and use, as the analogue of the Born–von Kármán boundary conditions, the conditions

$$\{E\,|\,MN_i\mathbf{t_i}\}\psi(\mathbf{r}) = \psi(\mathbf{r}) \quad (5.18)$$

$(i = 1, 2, 3)$. In the zero-field case eqn. (5.18) reduces to the Born–von Kármán boundary conditions. There is an important difference between the situations with and without magnetic fields. In the zero-field case, if one of the eigenfunctions $\psi(\mathbf{r})$ goes into itself under a translation $\{E\,|\,l_1N_1\mathbf{t_1} + l_2N_2\mathbf{t_2} + l_3N_3\mathbf{t_3}\}$ (l_1, l_2, l_3 integers) then so do all the functions $\psi(\mathbf{r} - \mathbf{T_n})$, where $\mathbf{T_n}$ is any lattice vector. Let us consider what happens for the magnetic translation operations. Suppose that we define a function

$$\phi_\mathbf{m}(\mathbf{r}) = \{E\,|\,M\mathbf{T_m}\}\psi(\mathbf{r}) \quad (5.19)$$

which is generated from $\psi(\mathbf{r})$ by a magnetic translation operation, then (using eqn. (5.16))

$$\{E\,|\,M(l_1N_1\mathbf{t_1} + l_2N_2\mathbf{t_2} + l_3N_3\mathbf{t_3})\}\phi_\mathbf{m}(\mathbf{r})$$
$$= \exp\left[-i[(l_1N_1\mathbf{t_1} + l_2N_2\mathbf{t_2} + l_3N_3\mathbf{t_3}) \wedge \mathbf{T_m}] \cdot \boldsymbol{\beta}\right]\phi_\mathbf{m}(\mathbf{r}). \quad (5.20)$$

Thus the boundary conditions in eqn. (5.18) can only be satisfied for all functions if the exponential factor in eqn. (5.20) is equal to unity, that is if

$$[(l_1N_1\mathbf{t_1} + l_2N_2\mathbf{t_2} + l_3N_3\mathbf{t_3}) \wedge \mathbf{T_m}] \cdot \boldsymbol{\beta} = 2\pi p \quad (5.21)$$

where p is an integer. If we write $\boldsymbol{\beta}$ and $\mathbf{T_m}$ in the form

$$\boldsymbol{\beta} = \lambda(n_1\mathbf{t_1} + n_2\mathbf{t_2} + n_3\mathbf{t_3}) \quad (5.22)$$

and

$$\mathbf{T_m} = m_1\mathbf{t_1} + m_2\mathbf{t_2} + m_3\mathbf{t_3} \quad (5.23)$$

then eqn. (5.21) gives

$$\lambda[(l_1 N_1 \mathbf{t}_1 + l_2 N_2 \mathbf{t}_2 + l_3 N_3 \mathbf{t}_3) \wedge (m_1 \mathbf{t}_1 + m_2 \mathbf{t}_2 + m_3 \mathbf{t}_3)]$$
$$\times (n_1 \mathbf{t}_1 + n_2 \mathbf{t}_2 + n_3 \mathbf{t}_3) = 2\pi p \qquad (5.24)$$

or

$$\lambda = 2\pi p / \Omega N \qquad (5.25)$$

where Ω is the volume of a unit cell ($= (\mathbf{t}_1 \wedge \mathbf{t}_2) \cdot \mathbf{t}_3$) and

$$N = (l_1 N_1 m_2 - l_2 N_2 m_1)n_3 + (l_2 N_2 m_3 - l_3 N_3 m_2)n_1 + (l_3 N_3 m_1 - l_1 N_1 m_3)n_2. \qquad (5.26)$$

Recalling that

$$\boldsymbol{\beta} = e\mathbf{B}/\hbar \qquad (5.27)$$

we see that eqn. (5.21) represents the imposition of restrictions on the external magnetic field for which it is possible to satisfy the generalization of the Born–von Kármán boundary conditions in eqn. (5.18), namely

$$\mathbf{B} = (hp/eN\Omega)(n_1 \mathbf{t}_1 + n_2 \mathbf{t}_2 + n_3 \mathbf{t}_3). \qquad (5.28)$$

To satisfy eqn. (5.21) we see that eqn. (5.28) implies that $\mathbf{B}$ must be parallel to a lattice vector $(n_1 \mathbf{t}_1 + n_2 \mathbf{t}_2 + n_3 \mathbf{t}_3)$ and its magnitude is quantized.

The form of the quantization condition imposed on $\mathbf{B}$ by the requirement to satisfy the generalized Born–von Kármán periodic boundary conditions is, however, rather curious. It is perhaps appropriate to try to see whether it is likely to lead to any physically observable effects. Quantum oscillations have been observed in numerous physical quantities for a metal or semi-conductor ; these oscillations are associated with the changes which occur in the positions of the Landau levels, relative to the Fermi energy, as the magnetic field is varied. These oscillations are found to be periodic in $1/|\mathbf{B}|$. Any oscillatory phenomenon that may be associated with the quantization condition in eqn. (5.28) will be much more complicated than this because of the appearance of both p and N in eqn. (5.28). p is any integer but N is the complicated sum of products of the 12 integers (l_i, m_i, n_i, N_i, where $i = 1, 2, 3$). N_1, N_2 and N_3 are taken to be large, as is usual in the case of the Born–von Kármán boundary conditions, and therefore the separation between adjacent allowed vectors for $\mathbf{B}$ is infinitesimally small. Therefore, it would seem that, although strictly speaking $\mathbf{B}$ is quantized, $\mathbf{B}$ can be regarded as being quasi-continuous in practice and physical effects associated with the quantization of $\mathbf{B}$ are unlikely to be observable.

The ideas outlined above for a crystal in an external magnetic field have been extended by Ashby and Miller (1965) to include a uniform external electric field as well.

§ 6. Particles and quasiparticles in crystalline solids

6.1. Electronic band structures

In § 2.2 we have already mentioned the important rôle which group theory has played in simplifying band-structure calculations for special wave vectors ; this depended on the use of projection operator techniques in the construction of symmetry-adapted functions for the various representations $\Gamma_j^{\,\mathbf{k}}$ for points of symmetry and lines of symmetry. As a result of increases in the size and speed of computing machinery it is now possible to calculate directly the energy $E_j(\mathbf{k})$ of an electron with an arbitrary wave vector $\mathbf{k}$ in a crystalline solid ; therefore the use of group theory in calculations of the energy bands $E_j(\mathbf{k})$ is fast becoming obsolete. Nevertheless the use of the group-theoretical labels for the energy bands is still convenient and will probably survive, so that it may be profitable to consider the question of the assignment of these labels on a slightly more physical basis than is commonly adopted. In the past, when using group theory in band-structure calculations for a given special point of symmetry, $\mathbf{k}$, it has been common to construct symmetry-adapted forms of $\psi_j(\mathbf{k}, \mathbf{r})$ for all the irreducible representations of the group of $\mathbf{k}$. These functions were then all used in turn in solving eqn. (2.10) and, in general, band theoreticians have paid little attention to the physical origins of the energy bands. There are two rather different approaches which can be adopted. The first is to use the free-electron, or ' empty-lattice ', model and the second is to consider the energy bands in relation to the electronic energy levels of the atoms of which the material is composed. In the past some economy could have been achieved by performing the calculations only for those representations which are compatible with the symmetries of those atomic states from which the bands in question are derived. Very often solutions of Schrödinger's equation were obtained only for the special points of symmetry in the Brillouin zone. The bands for general $\mathbf{k}$ were then obtained by some interpolation procedure, such as that devised by Slater and Koster (1954). Although the interpolation scheme devised by Slater and Koster was based on physical principles, other interpolation schemes have also been used, based on geometrical rather than physical principles. In such geometrical interpolation schemes the known point-group symmetry of the Brillouin zone, and therefore of the functions $E_j(\mathbf{k})$, could be exploited group-theoretically ; we shall return to the question of the symmetry of the $E_j(\mathbf{k})$ curves, rather than of the wave functions $\psi_j(\mathbf{k}, \mathbf{r})$, in § 6.2. Quite often, however, instead of using a numerical interpolation procedure the complete energy bands were simply sketched in, using the compatibility tables and assuming, based on the theorem due to Herring (1937) about accidental degeneracies, that bands belonging to the same irreducible representation along lines of symmetry will not intersect. When it is desired to determine the group-theoretical labels for the energy bands, without having used symmetry-adapted functions in one's band-structure calculations, it is even more important to consider the relationship of the energy bands of the solid either to the free-electron energy bands or to the electronic energy levels of free atoms.

In the ' free-electron ', or ' empty-lattice ', approach which was originally due to Shockley (1937) an electron in a solid is assumed to move in a potential which possesses the full translational symmetry of the crystal. The wave function of the electron can then be written as a Bloch function

$$\psi_j(\mathbf{k}, \mathbf{r}) = \exp{(i\mathbf{k} \cdot \mathbf{r})}u_j(\mathbf{k}, \mathbf{r}). \tag{6.1}$$

This Bloch function can be substituted into eqn. (2.10) to give

$$-\frac{\hbar^2}{2m}\{-\mathbf{k}^2 u_j(\mathbf{k}, \mathbf{r}) + 2i\mathbf{k} \cdot \nabla u_j(\mathbf{k}, \mathbf{r}) + \nabla^2 u_j(\mathbf{k}, \mathbf{r})\}$$
$$+ V(\mathbf{r})u_j(\mathbf{k}, \mathbf{r}) = E_j(\mathbf{k})u_j(\mathbf{k}, \mathbf{r}) \tag{6.2}$$

or

$$\nabla^2 u_j(\mathbf{k}, \mathbf{r}) + 2i\mathbf{k} \cdot \nabla u_j(\mathbf{k}, \mathbf{r}) + \frac{2m}{\hbar^2} \left\{ E_j(\mathbf{k}) - \frac{\hbar^2 \mathbf{k}^2}{2m} - V(\mathbf{r}) \right\} u_j(\mathbf{k}, \mathbf{r}) = 0. \qquad (6.3)$$

In any real metal it is difficult to determine a reliable expression for $V(\mathbf{r})$, but if the crystal potential $V(\mathbf{r})$ is set equal to zero throughout the crystal it is possible to solve eqn. (6.3) analytically. This is the free-electron approximation, or the empty-lattice model, and it is a valuable approach because, for many of the simple metals, it gives a good first approximation to their electronic band structures. A fairly detailed account of free-electron band structures is given in the book by Jones (1960). When $V(\mathbf{r})$ is set equal to zero the solutions of eqn. (6.3) take the form

$$u_j(\mathbf{k}, \mathbf{r}) = \exp\left(i\mathbf{G_n} \cdot \mathbf{r}\right) \qquad (6.4)$$

where $\mathbf{G_n}$ is a reciprocal lattice vector given by

$$\mathbf{G_n} = n_1 \mathbf{g_1} + n_2 \mathbf{g_2} + n_3 \mathbf{g_3} \qquad (6.5)$$

where n_1, n_2 and n_3 are integers. The corresponding energies $E_j(\mathbf{k})$ are given by

$$E_j(\mathbf{k}) = \frac{\hbar^2}{2m} \left| \mathbf{k} + \mathbf{G_n} \right|^2. \qquad (6.6)$$

With the form for $u_j(\mathbf{k}, \mathbf{r})$ given in eqn. (6.4) the corresponding expression for $\psi_j(\mathbf{k}, \mathbf{r})$ is given by

$$\psi_j(\mathbf{k}, \mathbf{r}) = \exp\left\{i(\mathbf{k} + \mathbf{G_n}) \cdot \mathbf{r}\right\} \qquad (6.7)$$

which is, of course, just a plane wave. If we regard $\mathbf{k}$ as being restricted to end within the first Brillouin zone, then for any given $\mathbf{k}$ one simply has to substitute all the possible values of $\mathbf{G_n}$ into eqn. (6.6) to determine the energies of all the bands at $\mathbf{k}$. The different energy bands at $\mathbf{k}$ can be assigned to the various irreducible representations of $\mathbf{G^k}$, the little group at $\mathbf{k}$ or the group of the wave vector $\mathbf{k}$, by studying the transformation properties of $\psi_j(\mathbf{k}, \mathbf{r})$ in eqn. (6.7) under the symmetry operations of $\mathbf{G^k}$.

The details of the determination of the symmetry assignments of the free-electron bands for the body-centred cubic and face-centred cubic structures have been given in the books by Jones (1960) and Cracknell and Wong (1973). The free-electron energy bands for a b.c.c. metal with a lattice constant appropriate to Na are shown in fig. 9 (b). If a rather small, but non-zero, value of $V(\mathbf{r})$ is now introduced throughout the crystal the energy bands will be only slightly perturbed from the free-electron energy bands ; this is illustrated for the case of Na in fig. 9 (a). Although it may change the shapes of the bands and may also lift some of the degeneracies that were present in the free-electron bands, the introduction of a non-zero $V(\mathbf{r})$, which still possesses the full symmetry of the crystal, will not alter the symmetries of the wave functions, that is it will not alter the group-theoretical labels that we assign to the energy bands. Thus for a simple metal it is often possible to establish the group-theoretical labels for many of the energy bands by inspection from a comparison with the free-electron energy bands for that structure, see fig. 9. However, for a more complicated metal, such as a transition metal or a rare-earth metal, the band structure may be so much perturbed from the free-electron band structure that the approach just described may be of little use.

In the alternative approach to the determination of group-theoretical labels for electronic energy bands in crystalline solids we seek to determine the relationship that exists between these energy bands for the solid and the electronic energy levels of the free atoms of the elements of which the solid is made (Sedaghat and Cracknell 1974). This can be done most easily by making use of tight-binding wave functions which we write in the form

$$\phi_j(\mathbf{k}, \mathbf{r}) = N^{-1/2} \exp(i\mathbf{k} \cdot \mathbf{r}) \sum_n w_j(\mathbf{r} - \mathbf{R}_n) h(\mathbf{r}, \mathbf{R}_n) \tag{6.8}$$

where N is the number of atoms in the crystal, $w_j(\mathbf{r})$ is a wave function for an electron in a free atom, $\mathbf{R}_n$ are the position vectors of the nuclei of the atoms, the summation over n is over all the atoms in the specimen, and $h(\mathbf{r}, \mathbf{R}_n)$ is a function, somewhat akin to a delta function, defined by

$$h(\mathbf{r}, \mathbf{R}_n) = 1 \tag{6.9}$$

for the site of the nucleus $\mathbf{R}_n$ which is closest to the electron and

$$h(\mathbf{r}, \mathbf{R}_n) = 0 \tag{6.10}$$

for all other $\mathbf{R}_n$. We assume that we are concerned with a crystal with only one atom per unit cell. Suppose that we imagine the atoms to be arranged in the crystallographic structure which is actually observed to occur in this particular metallic element in solid form, but with a variable value of the lattice constant a. If a is very large the energies of the $3d$ and $4s$ electrons will be perturbed very little from their values in the free atom. If a is imagined to be reduced steadily the energy bands will widen and may overlap and mix with one another. However, because the symmetry (that is the space group $\mathbf{G}$) is not changed as a is varied, the group-theoretical label $\Gamma_j^{\mathbf{k}}$ which is the label of that irreducible representation of $\mathbf{G}$ to which $\psi_j(\mathbf{k}, \mathbf{r})$ belongs, also cannot change as a is varied; this is because we expect $\psi_j(\mathbf{k}, \mathbf{r})$ to vary continuously with a. This 'adiabatic' principle can therefore be used to determine the symmetries of the wave functions $\psi_j(\mathbf{k}, \mathbf{r})$ for the bands in the solid with its true value of a.

The tight-binding wave function $\phi_j(\mathbf{k}, \mathbf{r})$ in eqn. (6.8) is of the form in eqn. (6.1) where

$$u_j(\mathbf{k}, \mathbf{r}) = N^{-1/2} \sum_n w_j(\mathbf{r} - \mathbf{R}_n) h(\mathbf{r}, \mathbf{R}_n). \tag{6.11}$$

$u_j(\mathbf{k}, \mathbf{r})$ is required, by Bloch's theorem, to possess the full translational symmetry of the Bravais lattice of the crystal; in group-theoretical terms this means that $u_j(\mathbf{k}, \mathbf{r})$ must belong to one of the space-group representations at Γ, that is at $\mathbf{k} = \mathbf{0}$. Which of the $\mathbf{k} = \mathbf{0}$ representations $u_j(\mathbf{k}, \mathbf{r})$ actually belongs to, let us call it Γ_p^Γ, will be determined by the rotational symmetry properties of $w_j(\mathbf{r})$, that is whether $w_j(\mathbf{r})$ is an s, p, d or f function. This can be ascertained from tables of the compatibility relations between the irreducible representations of $O(3)$, the group of the symmetry operations of a free atom, and the irreducible representations of the point group of the crystal. These tables are given, for example, by Koster *et al.* (1963) and an extract from them, for the point group 432 (O), has been reproduced already in table 2.

The symmetry of the wave function $\phi_j(\mathbf{k}, \mathbf{r})$ can be determined by study-ing the effects of the various space-group operations $\{S|\mathbf{w}\}$ on $\phi_j(\mathbf{k}, \mathbf{r})$. We have already seen that the function $N^{-1/2} \sum_n w_j(\mathbf{r} - \mathbf{R}_n) h(\mathbf{r}, \mathbf{R}_n)$ belongs to Γ_p^{Γ}, one of the $\mathbf{k} = \mathbf{0}$ representations. The identification of $\Gamma_q^{\mathbf{k}}$, the space-group representation to which $\exp(i\mathbf{k} \cdot \mathbf{r})$ belongs, can be determined by considering

$$\{S|\mathbf{w}\} \exp(i\mathbf{k} \cdot \mathbf{r}) = \exp\{i\mathbf{k} \cdot S^{-1}(\mathbf{r} - \mathbf{w})\} = \exp\{iS\mathbf{k} \cdot (\mathbf{r} - \mathbf{w})\} \quad (6.12)$$

(see eqn. (3.6.4) of Bradley and Cracknell (1972)). Therefore

$$\{S|\mathbf{w}\} \exp(i\mathbf{k} \cdot \mathbf{r}) = \exp(-iS\mathbf{k} \cdot \mathbf{w}) \exp(iS\mathbf{k} \cdot \mathbf{r}). \quad (6.13)$$

Since we restrict $\{S|\mathbf{w}\}$ to those operations which are in $\mathbf{G}^{\mathbf{k}}$, the group of $\mathbf{k}$, this means that $S\mathbf{k} \equiv \mathbf{k}$ and therefore $\mathbf{D}_q^{\mathbf{k}}(\{S|\mathbf{w}\})$, the matrix representative of $\{S|\mathbf{w}\}$ in the representation $\Gamma_q^{\mathbf{k}}$ to which $\exp(i\mathbf{k} \cdot \mathbf{r})$ belongs, is given by

$$\mathbf{D}_q^{\mathbf{k}}(\{S|\mathbf{w}\}) = \exp(-iS\mathbf{k} \cdot \mathbf{w}) \quad (6.14)$$

and the use of this equation enables $\Gamma_q^{\mathbf{k}}$ to be identified. The symmetry of $\phi_j(\mathbf{k}, \mathbf{r})$ can then be determined from the reduction of the Kronecker product

$$\Gamma_p^{\Gamma} \boxtimes \Gamma_q^{\mathbf{k}} = \sum_r c_{pq, r} \Gamma_r^{\mathbf{k}}. \quad (6.15)$$

For Γ, the point at the centre of the Brillouin zone, $\mathbf{k} = \mathbf{0}$ and then eqn. (6.14) leads to the simple results which we have already noted intuitively both for Γ and for those lines of symmetry which pass through Γ. For a symmorphic space group $\mathbf{w}$ can always be chosen to be zero so that $\Gamma_q^{\mathbf{k}}$ is the identity representation of $\mathbf{G}^{\mathbf{k}}$.

The procedure described above has been used by Sedaghat and Cracknell (1974) to determine the symmetries of the energy bands derived from the various atomic energy levels at the special points of symmetry in the f.c.c. and b.c.c. structures. The results are reproduced in table 20. Consider, for example, the bands of metallic Na which has the b.c.c. structure, see fig. 9. The electronic structure of a Na atom is $1s^2 2s^2 2p^6 3s^1$. By comparison of table 20 (a) with the labels in fig. 9 it can be seen that the lowest conduction band in Na can be regarded as a $3s$ band but that considerable overlap with the $3p$ band occurs at the edges of the Brillouin zone, namely at N and P. Somewhat surprisingly, it even overlaps with the $3d$ band at H. Among the higher conduction bands we see that a very considerable overlap between the $3p$ and $3d$ bands is apparent in this calculated band structure. It should perhaps be emphasized that although we describe these bands as originating from $3s$, $3p$ or $3d$ atomic energy levels, by a continuous evolution as the potential changes when the atoms are brought closer together, we do *not* imply that $l = 0$ and $l = 1$ survive as good quantum numbers for these bands. In terms of wave functions, therefore, we do not imply that only spherical harmonics with $l = 0$ and $l = 1$, respectively, would appear in the wave func-tions for these bands ; all that we imply is that the spherical harmonics with other values of l can only enter the expansions of the wave functions provided they occur in combinations with the symmetries specified in table 20 (a). An extensive discussion of the generation of such symmetry-adapted combina-tions of spherical harmonics will be found, for example, in chapter 2 of the book by Bradley and Cracknell (1972).

Instead of studying the transformation properties of $\phi_j(\mathbf{k}, \mathbf{r})$ under the symmetry operations $\{S|\mathbf{w}\}$ it would also be possible to make use of the tabulated sets of basis functions which are available for these space groups (see table IV of Altmann and Cracknell (1965)) and determine the symmetries of the bands derived from the various atomic energy levels by identifying those representations which have bases involving the appropriate values of l. In the case of the h.c.p. structure, because there are two atoms in the unit cell and because the symmetry of the structure is described by a non-symmorphic space group, it is very much easier to determine the symmetries of s, p, d and f bands using the published tables of basis functions given by Altmann and Bradley (1965), instead of constructing the corresponding functions similar to $\phi_j(\mathbf{k}, \mathbf{r})$ and then studying their transformation properties. The results are given in table 21. In tables 20 and 21 we have only given the results explicitly at the points of symmetry in the Brillouin zone. The symmetries of the energy bands along the lines of symmetry can be determined using the compatibility tables which are given by Bouckaert *et al.* (1936) for the f.c.c. and b.c.c. structures and by Altmann and Bradley (1965) for the h.c.p. structure.

In the above discussion we have ignored the existence of spin when considering the assignment of group-theoretical labels to the electronic energy

Table 20. Symmetries of bands at special points of symmetry for b.c.c. and f.c.c. structures

(a) b.c.c.

	Γ	H	N	P
s	Γ_1	H_1	N_1	P_1
p	Γ_{15}	H_{15}	$N_1{}', N_3{}', N_4{}'$	P_4
d	$\Gamma_{12},$ $\Gamma_{25}{}'$	$H_{12},$ $H_{25}{}'$	N_1, N_4 N_1, N_2, N_3	P_3 P_4
f	$\Gamma_2{}'$ Γ_{15} Γ_{25}	$H_2{}'$ H_{15} H_{25}	$N_3{}'$ $N_1{}', N_3{}', N_4{}'$ $N_1{}', N_2{}', N_4{}'$	P_1 P_4 P_5
	Γ	H	N	P
s	$\Gamma_1{}^+$	$H_1{}^+$	$N_1{}^+$	P_1
p	$\Gamma_4{}^-$	$H_4{}^-$	$N_2{}^-, N_3{}^-, N_4{}^-$	P_4
d	$\Gamma_3{}^+$ $\Gamma_5{}^+$	$H_3{}^+$ $H_5{}^+$	$N_1{}^+, N_2{}^+$ $N_1{}^+, N_3{}^+, N_4{}^+$	P_3 P_4
f	$\Gamma_2{}^-$ $\Gamma_4{}^-$ $\Gamma_5{}^-$	$H_2{}^-$ $H_4{}^-$ $H_5{}^-$	$N_2{}^-$ $N_2{}^-, N_3{}^-, N_4{}^-$ $N_1{}^-, N_3{}^-, N_4{}^-$	P_1 P_4 P_5

Table 20 (*Continued*)

(*b*) f.c.c.

	Γ	X	L	W
s	Γ_1	X_1	L_1	W_1
p	Γ_{15}	$X_4{}', X_5{}'$	$L_2{}', L_3{}'$	$W_2{}', W_3$
d	Γ_{12} $\Gamma_{25}{}'$	X_1, X_2 X_3, X_5	L_3 L_1, L_3	$W_1, W_2{}'$ $W_1{}', W_3$
f	$\Gamma_2{}'$ Γ_{15} Γ_{25}	$X_2{}'$ $X_4{}', X_5{}'$ $X_3{}', X_5{}'$	$L_2{}'$ $L_2{}', L_3{}'$ $L_1{}', L_3{}'$	W_2 $W_2{}', W_3$ W_1, W_3
	Γ	X	L	W
s	$\Gamma_1{}^+$	$X_1{}^+$	$L_1{}^+$	W_1
p	$\Gamma_4{}^-$	$X_3{}^-, X_5{}^-$	$L_2{}^-, L_3{}^-$	W_2, W_5
d	$\Gamma_3{}^+$ $\Gamma_5{}^+$	$X_1{}^+, X_2{}^+$ $X_4{}^+, X_5{}^+$	$L_3{}^+$ $L_1{}^+, L_3{}^+$	W_1, W_2 W_4, W_5
f	$\Gamma_2{}^-$ $\Gamma_4{}^-$ $\Gamma_5{}^-$	$X_2{}^-$ $X_3{}^-, X_5{}^-$ $X_4{}^-, X_5{}^-$	$L_2{}^-$ $L_2{}^-, L_3{}^-$ $L_1{}^-, L_3{}^-$	W_3 W_2, W_5 W_1, W_5

Note. For each structure the results are given first in the notation of Bouckaert *et al.* (1936) and then in the notation of Miller and Love (1967). The results for the lines of symmetry can be obtained from this table by using the compatibilities given by Bouckaert *et al.* (1936) ; for the f.c.c. structure these compatibilities have also been reproduced in table 5. (Sedaghat and Cracknell 1974.)

bands of a crystalline solid. When spin is included we find that, as before, the tight-binding wave function $u_j(\mathbf{k}, \mathbf{r})$ in eqn. (6.11) will belong to $\Gamma_p{}^\Gamma$, which is one of the space-group representations at $\mathbf{k} = \mathbf{0}$ in the Brillouin zone ; but now $\Gamma_p{}^\Gamma$ will be a double-valued representation instead of a single-valued representation. The method for the identification of the group-theoretical labels for the energy bands when spin is included consists of a relatively simple adaptation of the method we have already described. The method has been considered by Sedaghat and Cracknell (1974) who have also worked out the details for b.c.c., f.c.c. and h.c.p. metals.

In the examples considered so far we have been concerned with materials that do not exhibit spontaneous magnetic ordering. The method which we have described for single-valued representations and which we have indicated for double-valued representations can readily be adapted to the case of a material which exhibits ferromagnetic or simple antiferromagnetic ordering. The symmetry of such a material, for example a ferromagnetic metal (Falicov and Ruvalds 1968, Cracknell 1969 c, 1970 a, b) or an insulating antiferromagnetic material such as MnF_2 (Dimmock and Wheeler 1962 a) has to be described by the appropriate magnetic space group **M** which will, in general, contain fewer symmetry operations than the (grey) space group, **G₀**, which describes the symmetry of the atomic positions in the crystal when magnetic ordering is neglected. For a ferromagnetic metal the direction of the magnetization **M**, relative to the crystallographic axes can be altered by the application of a relatively small external magnetic field. The point-group symmetry of a single domain of a ferromagnetic metal will be given by the

Table 21. Symmetries of bands at special points of symmetry for h.c.p. structure

	Γ
s	$A_{1+}{}',\ A_{1-}{}'$
p	$A_{2+}{}'',\ E_+{}',\ A_{2-}{}'',\ E_-{}'$
d	$A_{1+}{}',\ E_+{}',\ E_+{}'',\ A_{1-}{}',\ E_-{}',\ E_-{}''$
f	$A_{1+}{}',\ A_{2+}{}',\ A_{2+}{}'',\ E_+{}',\ E_+{}'',\ A_{1-}{}',\ A_{2-}{}',\ A_{2-}{}'',\ E_-{}',\ E_-{}''$

	M
s	$\mathscr{A}_+{}',\ \mathscr{A}_-{}'$
p	$\mathscr{A}_+{}',\ \mathscr{B}_+{}',\ \mathscr{B}_+{}'',\ \mathscr{A}_-{}',\ \mathscr{B}_-{}',\ \mathscr{B}_-{}''$
d	$2\mathscr{A}_+{}',\ \mathscr{A}_+{}'',\ \mathscr{B}_+{}',\ \mathscr{B}_+{}'',\ 2\mathscr{A}_-{}',\ \mathscr{A}_-{}'',\ \mathscr{B}_-{}',\ \mathscr{B}_-{}''$
f	$2\mathscr{A}_+{}',\ \mathscr{A}_+{}'',\ 2\mathscr{B}_+{}',\ 2\mathscr{B}_+{}'',\ 2\mathscr{A}_-{}',\ \mathscr{A}_-{}'',\ 2\mathscr{B}_-{}',\ 2\mathscr{B}_-{}''$

	A
s	$A_1{}^{(2)}$
p	$A_1{}^{(2)},\ E^{(4)}$
d	$A_1{}^{(2)},\ 2E^{(4)}$
f	$2A_1{}^{(2)},\ A_2{}^{(2)},\ 2E^{(4)}$

	L
s	$\mathscr{E}_1$
p	$2\mathscr{E}_1,\ \mathscr{E}_2$
d	$3\mathscr{E}_1,\ 2\mathscr{E}_2$
f	$4\mathscr{E}_1,\ 3\mathscr{E}_2$

	K
s	E'
p	$A_1{}',\ A_2{}',\ E',\ E''$
d	$A_1{}',\ A_1{}'',\ A_2{}',\ A_2{}'',\ 2E',\ E''$
f	$A_1{}',\ A_2{}',\ A_1{}'',\ A_2{}'',\ 3E',\ 2E''$

	H
s	$E*$
p	$A^{(2)},\ 2E$
d	$2A^{(2)},\ E,\ 2E*$
f	$2A^{(2)},\ 2E,\ 3E*$

Note. The representations are labelled in the notation of Altmann and Bradley (1965) who also give the correspondence between their labels and those used by Herring (1942). (Sedaghat and Cracknell 1974.)

intersection of the point group of the non-magnetic metal with the non-crystallographic Heesch–Shubnikov point group ∞/mm'. The corresponding space groups for ferromagnetic b.c.c., f.c.c. and h.c.p. metals magnetized parallel to important crystallographic directions are identified in table 22. The use of the appropriate magnetic space group will lead to a different scheme for labelling the energy bands and in many cases will lead to the prediction of fewer essential degeneracies than occur in the band structure of the non-magnetic phase of the material with the same crystallographic structure. This re-labelling of the bands in the magnetic phase may be performed

Table 22.   Heesch–Shubnikov space groups for ferromagnetic b.c.c., f.c.c. and h.c.p. metals

b.c.c.	**M**		
	[001]	$I4/mm'm'$	$\text{Ш}_{139}{}^{537}$
	[111]	$R\bar{3}m'$	$\text{Ш}_{166}{}^{101}$
	[110]	$Fm'm'm$	$\text{Ш}_{69}{}^{524}$
f.c.c.	**M**		
	[001]	$I4/mm'm'$	$\text{Ш}_{139}{}^{537}$
	[111]	$R\bar{3}m'$	$\text{Ш}_{166}{}^{101}$
	[110]	$Im'm'm$	$\text{Ш}_{71}{}^{536}$
h.c.p.	**M**		
	[0001]	$P6_3/mm'c'$	$\text{Ш}_{194}{}^{270}$
	[10$\bar{1}$0]	$Cm'cm'$	$\text{Ш}_{63}{}^{461}$
	[11$\bar{2}$0]	$Cm'cm'$	$\text{Ш}_{63}{}^{463}$
	[$uv\dagger$0]	$P2_1'/m'$	$\text{Ш}_{11}{}^{54}$

Note.   For further details see § 2.2.1 of Cracknell (1974 a).

ab initio using eqn. (6.15) or, alternatively, by making use of the compatibilities determined by the subgroup relation which exists between **M** and $\mathbf{G}_0$. For more complicated forms of magnetic ordering it may no longer be possible to use one of the Heesch–Shubnikov space groups to describe the symmetry of the magnetic structure and the development of other more generalized symmetry groups may be necessary. When the crystallographic symmetry and the symmetry of the arrangement of the magnetic moments are incommensurate, the effect of the magnetic ordering on the band structure is complicated (Herring 1966, Cracknell 1971 a) and the assignment of group-theoretical labels may become difficult.

6.2. *The symmetry of the Fermi surface*

Whereas in §§ 2.2 and 6.1 we have been concerned with studying the symmetry properties of the wave functions which are the eigenfunctions of the Hamiltonian, $\mathcal{H}$, of an electron in a crystalline solid in the independent-particle approximation, we now turn to the consideration of the symmetry properties of the eigenvalues of $\mathcal{H}$, where these eigenvalues $E_j(\mathbf{k})$ are regarded as quasi-continuous functions of $\mathbf{k}$. Suppose that the symmetry of a crystalline specimen of a given solid is described by a space group $\mathbf{G}$ which is based on a Bravais lattice with primitive translations $\mathbf{t}_1$, $\mathbf{t}_2$ and $\mathbf{t}_3$. Since $E_j(\mathbf{k})$ is a function of the three-dimensional vector $\mathbf{k}$ we would need a four-dimensional space to give a complete representation of the function $E_j(\mathbf{k})$. To avoid this difficulty it is convenient to construct constant-energy surfaces defined by

$$E_j(\mathbf{k}) = E \qquad (6.16)$$

in the three-dimensional vector space of $\mathbf{k}$. In the case of a metallic solid there is one particular constant-energy surface which is of special importance, namely the Fermi surface for which the constant E is equal to E_F, the Fermi energy. We require to study the symmetry properties of the function $E_j(\mathbf{k})$ in the reciprocal space which is based on the primitive translations $\mathbf{g}_1$, $\mathbf{g}_2$ and $\mathbf{g}_3$ of the reciprocal lattice where $\mathbf{g}_i \cdot \mathbf{t}_j = 2\pi\delta_{ij}$.

Suppose that $\psi_j(\mathbf{k}, \mathbf{r})$ is the wave function corresponding to the energy $E_j(\mathbf{k})$ and that $\{R|\mathbf{v}\}$ is the Seitz space-group symbol for an arbitrary operation of the space group $\mathbf{G}$. Then using Wigner's theorem, which was mentioned previously in § 1.3, it follows that $\{R|\mathbf{v}\}\psi_j(\mathbf{k}, \mathbf{r})$ is also a solution of Schrödinger's equation for the solid and corresponds to the same energy $E_j(\mathbf{k})$. But

$$\{R|\mathbf{v}\}\psi_j(\mathbf{k}, \mathbf{r}) = \psi_j(R\mathbf{k}, \mathbf{r} - \mathbf{v}) \qquad (6.17)$$

(see, for example, eqn. (3.6.6) of Bradley and Cracknell (1972)) and this equation means that the spectrum of the energy eigenvalues $E_j(\mathbf{k})$ is the same at $\mathbf{k}$ and at $R\mathbf{k}$. We have already seen in § 5.2 that, in the absence of magnetic ordering, $E_j(\mathbf{k})$ and $E_j(-\mathbf{k})$ are degenerate. Therefore any constant-energy surface, and in particular the Fermi surface, must possess all the symmetry of some point group which we shall designate $\mathbf{P}'$, where we consider the constant-energy surfaces in terms of a Brillouin zone which is centred at Γ. If $\mathbf{P}$ is the isogonal point group[†] of $\mathbf{G}$ then

$$\mathbf{P}' = \mathbf{P} \qquad (6.18)$$

if $\mathbf{P}$ contains I, the operation of space inversion, or

$$\mathbf{P}' = \mathbf{P} \otimes \bar{I} \qquad (6.19)$$

if $\mathbf{P}$ does not contain I. If one regards the Brillouin zone as a unit cell in the repeated zone scheme then $E_j(\mathbf{k})$ will possess all the symmetry of the symmorphic space group $\mathbf{G}'$ which is obtained by associating the operations of the point group $\mathbf{P}'$ with the translations $(n_1\mathbf{g}_1 + n_2\mathbf{g}_2 + n_3\mathbf{g}_3)$ of the reciprocal

---

[†] The isogonal point group is constructed by taking the point-group operations that would be obtained by setting $\mathbf{v} = \mathbf{0}$ in all the operations of the space group $\mathbf{G}$.

lattice when n_1, n_2 and n_3 are integers. $\mathbf{G}'$ need not be the same space group as $\mathbf{G}$, the space group of the crystal, and indeed if $\mathbf{G}$ is a non-symmorphic space group, or if $\mathbf{G}$ does not contain the space inversion operation, $\mathbf{G}'$ cannot be the same as $\mathbf{G}$.   In practice it is quite straightforward to identify $\mathbf{G}'$ for any one of the 230 classical space groups, see table 23.   The above argument can easily be extended to the case of a magnetically ordered crystal or a crystal situated in an external magnetic field when the symmetry is described by one of the Heesch–Shubnikov space groups (for details see Cracknell (1973 b)).   It is this extension which is responsible for the manner in which the co-representation domain was defined in § 1.4.

Although it will always be the case that any given constant-energy surface, for either a classical space group or for a Heesch–Shubnikov space group, must possess the symmetry of the appropriate point group $\mathbf{P}'$ relative to the origin, Γ, of the reciprocal lattice, this may not always be the most convenient

Table 23.   The identification of $\mathbf{G}'$, the space group of $E_j(\mathbf{k})$

Crystal system	Laue group of $\mathbf{G}$	Bravais lattice and reciprocal lattice		$\mathbf{G}'$	
Triclinic	$\bar{1}$	P	P	$P\bar{1}$	2
Monoclinic	$2/m$	P	P	$P2/m$	10
		B	B	$B2/m$	12
Orthorhombic	mmm	P	P	$Pmmm$	47
		C	C	$Cmmm$	65
		I	F	$Fmmm$	69
		F	I	$Immm$	71
Tetragonal	$4/m$	P	P	$P4/m$	83
		I	I	$I4/m$	87
	$4/mmm$	P	P	$P4/mmm$	123
		I	I	$I4/mmm$	139
Trigonal	$\bar{3}$	P	P	$P\bar{3}$	147
		R	R	$R\bar{3}$	148
	$\bar{3}m$	P	P	$P\bar{3}1m$	162
		P	P	$P\bar{3}m1$	164
		R	R	$R\bar{3}m$	166
Hexagonal	$6/m$	P	P	$P6/m$	175
	$6/mmm$	P	P	$P6/mmm$	191
Cubic	$m3$	P	P	$Pm3$	200
		I	F	$Fm3$	202
		F	I	$Im3$	204
	$m3m$	P	P	$Pm3m$	221
		I	F	$Fm3m$	225
		F	I	$Im3m$	229

Adapted from Jan (1972).

way of looking at that particular constant-energy surface. Two examples of a hypothetical constant-energy surface in two dimensions are illustrated in fig. 15 for a plane space group with $\mathbf{P}' = 4/mmm$. The surface shown in fig. 15 (*a*) is centred at Γ and possesses the symmetry of $4/mmm$. However, while the system of pockets comprising the surface shown in fig. 15 (*b*) still possesses

Fig. 15

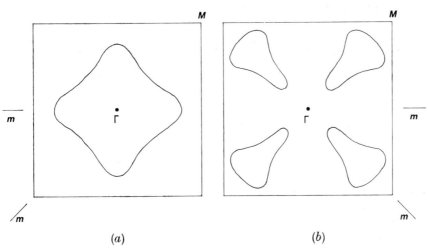

(*a*)   (*b*)

Two-dimensional illustrations of a Fermi surface with $\mathbf{P}' = 4/mmm$.

all the symmetry of the point group $4/mmm$ about Γ it is likely to be more convenient to consider just one of the pockets on its own. The point-group symmetry of one of these pockets will not in general be $\mathbf{P}'$ but will be some other point group, which we call $\mathbf{P}'(\mathbf{k}_0)$ and which remains to be identified. $\mathbf{k}_0$ is some wave vector which it seems convenient to choose as an origin when describing the point-group symmetry of one of these pockets. There are special problems associated with constant-energy surfaces or Fermi surfaces which intersect the Brillouin zone boundary. In such cases it may be possible to produce only closed surfaces by remembering that the Brillouin zone is repeated around each reciprocal-lattice point. Alternatively, one may have a multiply-connected constant-energy surface for which it is not possible, by any change of origin in the repeated zone scheme, to produce a set of closed surfaces.

Let us now see how to identify $\mathbf{P}'(\mathbf{k}_0)$. Suppose that $\mathbf{k}$ is a wave vector which terminates on a certain constant-energy surface

$$E_j(\mathbf{k}) = E \tag{6.20}$$

with a small closed pocket near $\mathbf{k}_0$. We then write $\mathbf{k}$ in the form

$$\mathbf{k} = \mathbf{k}_0 + \boldsymbol{\varkappa}. \tag{6.21}$$

From eqn. (6.17) and the use of Wigner's theorem we see that the point specified by $R\mathbf{k}$ will also be on the same constant energy surface. From eqn. (6.21) we can write

$$R\mathbf{k} = R\mathbf{k}_0 + R\boldsymbol{\varkappa}. \tag{6.22}$$

There are three possibilities which may arise :

(i) $R\mathbf{k}_0$ may be identical to $\mathbf{k}_0$, in which case $R\mathbf{k}$ and $\mathbf{k}$ are on the same pocket of the constant-energy surface,

(ii) $R\mathbf{k}_0$ may only differ from $\mathbf{k}_0$ by some translation $(n_1\mathbf{g}_1 + n_2\mathbf{g}_2 + n_3\mathbf{g}_3)$ of the reciprocal lattice, in which case $R\mathbf{k}$ is a point on a pocket that is equivalent to the pocket on which $\mathbf{k}$ is situated,

(iii) $R\mathbf{k}_0$ is not equivalent to $\mathbf{k}_0$ so that $R\mathbf{k}$ and $\mathbf{k}$ are on different pockets of the constant energy surface.

Therefore the pocket of constant-energy surface near to $\mathbf{k}_0$ will have the symmetry of the point group $\mathbf{P}'(\mathbf{k}_0)$ which consists of all the operations of $\mathbf{P}'$ that behave according to (i) or (ii) but not those that behave according to (iii). $\mathbf{P}'(\mathbf{k}_0)$ is a, possibly trivial, subgroup of the point group $\mathbf{P}'$. The identification of the point group $\mathbf{P}'(\mathbf{k}_0)$ for crystals with the symmetry of a classical space group or of a Heesch–Shubnikov space group is discussed in detail elsewhere and the results have been tabulated (Cracknell 1971 c, 1973 a, b).

One motive behind studying the symmetry of constant-energy surfaces, and in particular the Fermi surface, is connected with attempts to obtain parametric expressions for the Fermi surfaces of metals (Cracknell 1971 c, Jan 1972, Cracknell and Chia 1973). We refer particularly to geometrical parametrization schemes for a given constant-energy surface, rather than to interpolation schemes, such as that of Slater and Koster (1954) for example, which aim to reproduce the complete band structure and are based on particular methods of calculation of band structures. In a geometrical parametrization scheme one obtains formulae for k_F $(=|\mathbf{k}_\mathrm{F}|)$ in terms of the direction of $\mathbf{k}_\mathrm{F}$, where $\mathbf{k}_\mathrm{F}$ is a wave vector which terminates on the Fermi surface ; this can be done without having to reconstruct the whole band structure. Possible formulae which can be used can be derived from expansions of k_F or from expansions of $E_j(\mathbf{k}_\mathrm{F})$. If one writes $\mathbf{k}_\mathrm{F}$ in spherical polar coordinates as $(k_\mathrm{F}, \theta_\mathrm{F}, \phi_\mathrm{F})$ it is natural to expand k_F in terms of spherical harmonics so that

$$k_\mathrm{F} = \sum_{l,\,m} A_{lm} Y_l^m(\theta_\mathrm{F},\,\phi_\mathrm{F}). \qquad (6.23)$$

If one uses the cartesian coordinates $k_{\mathrm{F}x}$, $k_{\mathrm{F}y}$ and $k_{\mathrm{F}z}$ it is natural to expand k_F as a series of polynomials in $k_{\mathrm{F}x}$, $k_{\mathrm{F}y}$ and $k_{\mathrm{F}z}$; these polynomials may be spherical harmonics or some other polynomials. Alternatively we can expand $E_j(\mathbf{k})$ in terms of plane waves, which is really just a three-dimensional Fourier expansion of $E_j(\mathbf{k})$, and write

$$E_j(\mathbf{k})/E_\mathrm{F} = \sum_{n_1,\,n_2,\,n_3} C_{n_1 n_2 n_3} \exp\{i(n_1\mathbf{g}_1 + n_2\mathbf{g}_2 + n_3\mathbf{g}_3)\,.\,\mathbf{k}\} \qquad (6.24)$$

so that for the Fermi surface itself we have

$$\sum_{n_1,\,n_2,\,n_3} C_{n_1 n_2 n_3} \exp\{i(n_1\mathbf{g}_1 + n_2\mathbf{g}_2 + n_3\mathbf{g}_3)\,.\,\mathbf{k}_\mathrm{F}\} = 1. \qquad (6.25)$$

At present the general topological features of the Fermi surfaces of most of the metallic elements are already known, apart from a few exceptions principally among the lanthanide and actinide elements (see, for example, Cracknell

(1971 a)). In addition to this knowledge, there are already available very detailed measurements of the dimensions of various sheets of the Fermi surfaces of a number of metallic elements; it is likely that the next few years will see the collection of similar data for many of the other metallic elements. It therefore seems very likely that in the next few years we shall also see a considerable increase in the use of geometric parametrization schemes as a convenient method for the representation and storage of such data. Group theory can play an important part in the development of these parametrization schemes. By using the known space group $\mathbf{G}'$, which we have defined above and which describes the symmetry of the constant-energy surfaces, it is possible to simplify the forms of expansions such as those given in eqns. (6.23) and (6.25). This will reduce the number of unknown coefficients that have to be determined from the available data for the Fermi surface of the metal in question. Simplified forms of eqn. (6.23) for expansions in terms of spherical harmonics, about various different points in the Brillouin zone, can be obtained for b.c.c. and f.c.c. metals from the tables of Altmann and Cracknell (1965) and for h.c.p. metals from the tables of Altmann and Bradley (1965). Simplified forms of eqn. (6.25), for expansions about various different points in the Brillouin zone, for b.c.c. and f.c.c. metals are given in the tables of Chia *et al.* (1973); an extract from these tables for expansions about those special wave vectors $\mathbf{k}_0$ which are actually of interest in real b.c.c. metals is reproduced in table 24.

In addition to its use in connection with geometrical parametrization schemes for Fermi surfaces, the identification of $\mathbf{P}'(\mathbf{k}_0)$ will also be important in connection with essential stationary values of $E_j(\mathbf{k})$ and Van Hove singularities (see § 6.7).

Table 24. Expansions for $E_j(\varkappa)/E_F$ for b.c.c. metals

$\mathbf{k}_0$	Terms in expansion	Restrictions
Γ, H	$C_{p_1 p_2 p_3} \cos p_1 a \kappa_x \cos p_2 a \kappa_y \cos p_3 a \kappa_z$	†
P	$C_{p_1 p_2 p_3} \cos p_1 a \kappa_x \cos p_2 a \kappa_y \cos p_3 a \kappa_z$ $D_{p_1 p_2 p_3} \sin p_1 a \kappa_x \sin p_2 a \kappa_y \sin p_3 a \kappa_z$	† †
N	$B_{p_1 p_2 p_3} \sin p_1 a \kappa_x \sin p_2 a \kappa_y \cos p_3 a \kappa_z$ $C_{p_1 p_2 p_3} \cos p_1 a \kappa_x \cos p_2 a \kappa_y \cos p_3 a \kappa_z$	$B_{p_1 p_2 p_3} = B_{p_2 p_1 p_3}$ $C_{p_1 p_2 p_3} = C_{p_2 p_1 p_3}$
Δ	$C_{p_1 p_2 p_3} \cos p_1 a \kappa_x \cos p_2 a \kappa_y \cos p_3 a \kappa_z$ $D_{p_1 p_2 p_3} \cos p_1 a \kappa_x \sin p_2 a \kappa_y \cos p_3 a \kappa_z$	$C_{p_1 p_2 p_3} = C_{p_3 p_2 p_1}$ $D_{p_1 p_2 p_3} = D_{p_3 p_2 p_1}$

$p_1 = m_1 + m_4$, $p_2 = m_2 + m_4$, $p_3 = m_3 + m_4$, where m_1, m_2 and m_3 are positive integers and $m_4 = 0$ or $\frac{1}{2}$.

† Coefficients with permutations of p_1, p_2 and p_3 are equal.

We should perhaps clarify the relationship between $\mathbf{P}'(\mathbf{k}_0)$ and the little group $\mathbf{G}^{\mathbf{k}_0}$ which is used, for example, in connection with the symmetries of wave functions. $\mathbf{G}^{\mathbf{k}_0}$ is a space group which is a subgroup of the space group

G which describes the symmetry of a crystal in real space. On the other hand, **G'**, **P'** and **P'(k₀)** are groups which are concerned with describing the symmetry of constant-energy surfaces and therefore are groups of symmetry operations which act in reciprocal space, or **k** space, and not in real space. Thus it is not surprising that in general **G'** is not the same space group as **G**, the space group of the crystal. **G'** is always a symmorphic space group even if **G** is non-symmorphic. Moreover, **G'** is always a classical space group, even for crystals for which the symmetry of the crystal in real space is described by one of the black and white space groups, **M**. Let us consider an electronic state labelled by $\Gamma_j{}^k$ which has wave function $\psi_j(\mathbf{k}, \mathbf{r})$ with corresponding eigenvalue $E_j(\mathbf{k})$. The symmetry of $\psi_j(\mathbf{k}, \mathbf{r})$, as a function of **r**, is described by $\Gamma_j{}^k$ which is one of the irreducible representations of the little group **G**ᵏ. **G**ᵏ⁰ therefore gives information about the form of the wave function, in real space, and the degeneracies of the eigenvalues, for the particular wave vector $\mathbf{k} = \mathbf{k}_0$. **P'(k₀)** describes the symmetry, in **k** space, of a constant energy surface in the vicinity of $\mathbf{k}_0$.

6.3. *Phonon symmetries*

The introduction of the use of space-group representations in labelling the phonon dispersion relations of a crystal and in studying their degeneracies is much newer than in the work on electronic band structures (see, for example, Cowley (1964), Johnson and Loudon (1964), Poulet (1965), Maradudin and Vosko (1968), Warren (1968)). The early experimental measurements of phonon dispersion relations by means of neutron scattering were mostly performed for important directions in crystals of high symmetry. Each normal mode could then be described in a meaningful way as either longitudinal acoustic (LA), transverse acoustic (TA), longitudinal optic (LO), or transverse optic (TO). However, as experimental measurements were extended to directions of lower symmetry and to crystals with more complicated structures, it became apparent that this simple scheme was inadequate for classifying the normal modes of vibration in many crystals. The use of space-group representations in labelling the phonon dispersion relations is now well-established and has been used for a number of crystalline materials. Extensive discussions of the principles involved in the use of group theory in this connection will be found in the articles by Maradudin and Vosko (1968), Warren (1968), and Montgomery (1969), while a computer programme for performing the analysis for different crystal structures has been developed by Worlton and Warren (1972).

The question of the definition of longitudinal and transverse waves in complex crystals is not trivial and simple ideas based on structures like rocksalt and zincblende are inadequate for the general case. Suppose we have an acoustic wave with a wave vector **k** corresponding to a long but finite wavelength, that is **k** is small but non-zero. The centre of mass of each primitive cell is moving periodically in time, and if the displacement of the centre of mass is in a straight line parallel to **k** the mode is LA ; if it is in a straight line or an ellipse lying in a plane normal to **k** the mode is TA. If neither of these situations applies, the classification into LA and TA is not valid. A similar argument applies to infra-red active optic modes, if for the displacement of the centre of mass one substitutes the electric dipole moment of the primitive cell. It follows that non-primitive translations for space-group elements are irrelevant, as the phase shift of the wave from one cell to the next can be made arbitrarily small by allowing **k** to tend towards zero in a well-defined direction. (In the same way macroscopic tensors for a crystal are governed by its point group, not by its space group.) Suppose the group of the wave vector **k** is derived from elements of the form $\{R|\mathbf{v}\}$ in the Seitz notation. When one applies the operation $\{R|\mathbf{v}\}$ to the acoustic mode the displacement of the centre of mass suffers a rotation R; therefore the acoustic mode must transform under $\{R|\mathbf{v}\}$ in the same way that a Euclidean vector transforms under the point-group operation R, provided that we can neglect the small phase shift produced by **v**. The three acoustic modes of wave vector **k** are mutually orthogonal, and it follows that they transform like the three orthogonal components of a polar vector in Euclidean space. It should be noted that these arguments apply strictly only in the limit of long waves, where the acoustic frequencies are determined by the elastic constants. In the neighbourhood of the Brillouin zone boundary the distinction between TA and LA can be quite meaningless, and cases exist where a branch is LA near the zone centre, passes smoothly through the zone boundary and becomes TA near the centre of the next zone (see, for example, fig. 6 of Montgomery and Dolling (1972)).

The use of group theory in labelling the normal modes of vibration of a crystalline solid and in determining the forms of the eigenvectors for these normal modes can be regarded as an extension of earlier work, which has been used particularly by chemists, in connection with the vibrations of isolated molecules (see, for example, Bhagavantam and Venkatarayudu (1948), Heine (1960), Cracknell (1968 a)). The essential result that is invoked is enshrined in Wigner's theorem which implies that each eigenfunction of the Hamiltonian, $\mathscr{H}$, of a particle or quasiparticle in a crystalline solid belongs to one of the irreducible representations of the space group, **G**, of the solid. Instead of being the Hamiltonian that describes the motion of an itinerant electron in the solid, $\mathscr{H}$ is now the Hamiltonian that describes the lattice vibrations of the solid. Suppose that each fundamental unit cell of a certain crystal contains r atoms and that there are N of these unit cells in the crystal. A general displacement of the crystal will involve $3r$ co-ordinates per unit cell or $3rN$ coordinates for the whole crystal. If one wishes to apply space group ideas to a finite crystal one has to use cyclic (Born–von Kármán) boundary conditions, when the number of normal modes will be $3rN$. Each normal mode is characterized by a wave vector **k**, and they are distributed among the N discrete allowed wave vectors in the Brillouin zone so that for each **k** there are $3r$ normal modes. The assignment of the normal modes of vibration of a crystal to the various irreducible representations of the space group **G** of that crystal can be determined by studying the transformation properties, under the symmetry operations of **G**, of the vector Ξ with $3rN$ components which are the displacements $u_\alpha(l\kappa)$ of the atoms in the crystal. Fortunately it is adequate to consider a vector ξ in only $3r$ dimensions which describes the displacements of the atoms in a single unit cell. However, it has to be remembered that some of the symmetry operations of the space group **G** will move atoms out of this cell and into a different cell; this has to be allowed for in constructing the $3r$-dimensional matrices based on $\boldsymbol{\xi}$.

We suppose that $\{R|\mathbf{v}\}$ is an element of the space group $\mathbf{G}$ of the crystal and we consider the effect of $\{R|\mathbf{v}\}$ on the displacements $u_\alpha(l\kappa)\{=[u_1(l\kappa),$ $u_2(l\kappa),\ u_3(l\kappa)]\}$ of all the atoms in the crystal, where $u_\alpha(l\kappa)$ is the α Cartesian coordinate of the displacement of the atom labelled by κ in the unit cell labelled by l. From this information we shall be able to construct a $3r$-dimensional representation $\Gamma_{\boldsymbol{\xi}}{}^{\mathbf{k}}$ which is the analogue of the representation $\Gamma_{\boldsymbol{\xi}}$ which can be constructed for the displacements of the atoms in a vibrating molecule. The position vector of the atom labelled by l and κ is $\mathbf{u}(l\kappa)+\mathbf{r}(l\kappa)$, referred to the origin, where $\mathbf{r}(l\kappa)$ is the vector of the equilibrium position of the atom, referred to the origin. The effect of $\{R|\mathbf{v}\}$ on this vector can be determined in the usual way and gives

$$\{R|\mathbf{v}\}\{\mathbf{u}(l\kappa)+\mathbf{r}(l\kappa)\}=R\{\mathbf{u}(l\kappa)+\mathbf{r}(l\kappa)\}+\mathbf{v}$$
$$=R\mathbf{u}(l\kappa)+\{R\mathbf{r}(l\kappa)+\mathbf{v}\}. \qquad (6.26)$$

From the form of eqn. (6.26) we see that it describes the final position, after the performance of $\{R|\mathbf{v}\}$, for an atom on a site specified by l', κ' which is in general different from the original site, and where l' and κ' are defined by

$$\mathbf{r}(l'\kappa')=R\mathbf{r}(l\kappa)+\mathbf{v}=\{R|\mathbf{v}\}\mathbf{r}(l\kappa). \qquad (6.27)$$

Equation (6.26) can therefore be interpreted to give a relation between the final displacement of the atom $l'\kappa'$ and the initial displacement of the atom $l\kappa$

$$\mathbf{u}'(l'\kappa')=R\mathbf{u}(l\kappa) \qquad (6.28)$$

where l' and κ' are defined by

$$\mathbf{r}(l'\kappa')=\{R|\mathbf{v}\}\mathbf{r}(l\kappa). \qquad (6.29)$$

The equation of motion for the atom $l\kappa$ can be written as

$$M_\kappa\ddot{u}_\alpha(l\kappa)=-\sum_{l'}\sum_{\kappa'}\sum_\beta\Phi_{\alpha\beta}(l\kappa\ ;\ l'\kappa')u_\beta(l'\kappa') \qquad (6.30)$$

where $\Phi_{\alpha\beta}(l\kappa\ ;\ l'\kappa')$ are the force constants between atoms specified by $l\kappa$ and $l'\kappa'$. Equation (6.30) can be rewritten as

$$\ddot{w}_\alpha(l\kappa)=-\sum_{l'}\sum_{\kappa'}\sum_\beta\mathbf{D}_{\alpha\beta}(l\kappa\ ;\ l'\kappa')w_\beta(l'\kappa') \qquad (6.31)$$

where

$$w_\alpha(l\kappa)=(M_\kappa)^{1/2}u_\alpha(l\kappa) \qquad (6.32)$$

and $\mathbf{D}_{\alpha\beta}(l\kappa\ ;\ l'\kappa')$ is called the dynamical matrix.

An arbitrary displacement $u_\alpha(l\kappa)$ can be expanded in terms of normal-mode eigenvectors so that, from the conventional theory of lattice vibrations in the harmonic approximation, we have

$$u_\alpha(l\kappa)=(M_\kappa)^{-1/2}\sum_j\sum_{\mathbf{k}}e_\alpha(\kappa|\mathbf{k}j)Q_j(\mathbf{k})\exp\left[i\{\mathbf{k}\ .\ \mathbf{r}(l\kappa)-\omega_j(\mathbf{k})t\}\right]. \qquad (6.33)$$

We study the symmetries of the normal modes for each wave vector separately. For this purpose we suppose that the crystal is vibrating in such a manner that only the normal modes with the chosen wave vector $\mathbf{k}$ are involved. The components of the displacements of the atoms in this vibration will be denoted by $u_\alpha{}^k(l\kappa)$. A complete expansion of $u_\alpha{}^k(l\kappa)$ can be obtained by including only the normal modes with the given value of $\mathbf{k}$. Thus, in place of eqn. (6.33) we have

$$u_\alpha{}^k(l\kappa) = (M_\kappa)^{-1/2} \sum_j e_\alpha(\kappa \,|\, \mathbf{k}j) Q_j(\mathbf{k}) \exp\left[i\{\mathbf{k} \cdot \mathbf{r}(l\kappa) - \omega_j(\mathbf{k})t\}\right]. \tag{6.34}$$

Since only a single value of $\mathbf{k}$ is involved, it is possible to obtain a relationship between $\mathbf{u}^k(l\kappa')$ and $\mathbf{u}^k(l'\kappa')$, the displacements of atoms occupying the same sites in different unit cells, namely

$$\mathbf{u}^k(l'\kappa') = \exp\left[i\mathbf{k} \cdot \{\mathbf{r}(l'\kappa') - \mathbf{r}(l\kappa')\}\right]\mathbf{u}^k(l\kappa'). \tag{6.35}$$

Therefore, using eqn. (6.28) we have

$$R\mathbf{u}^k(l\kappa) = \exp\left[i\mathbf{k} \cdot \{\mathbf{r}(l'\kappa') - \mathbf{r}(l\kappa')\}\right]\mathbf{u}'^k(l\kappa'). \tag{6.36}$$

Table 25.   Phonon symmetries for a crystal with the NaCl structure

$\mathbf{k}$	Phonon symmetries	
Γ	$2\Gamma_4{}^-$ $2(3)$	$2\Gamma_{15}$ $2(3)$
X	$2X_2{}^-\oplus2X_5{}^-$ $2(1)\quad2(2)$	$2X_4{}'\oplus2X_5{}'$ $2(1)\quad2(2)$
L	$L_1{}^+\oplus L_2{}^-\oplus L_3{}^+\oplus L_3{}^-$ $1(1)\quad1(1)\quad1(2)\quad1(2)$	$L_1\ \oplus L_2{}'\ \oplus L_3\ \oplus L_3{}'$ $1(1)\quad1(1)\quad1(2)\quad1(2)$
W	$W_1\oplus W_4\oplus2W_5$ $1(1)\quad1(1)\quad2(2)$	$W_1\oplus W_2{}'\oplus2W_3$ $1(1)\quad1(1)\quad2(2)$
Δ	$2\Delta_1\oplus2\Delta_5$ $2(1)\quad2(2)$	$2\Delta_1\oplus2\Delta_5$ $2(1)\quad2(2)$
Λ	$2\Lambda_1\oplus2\Lambda_3$ $2(1)\quad2(2)$	$2\Lambda_1\oplus2\Lambda_3$ $2(1)\quad2(2)$
Σ	$2\Sigma_1\oplus2\Sigma_2\oplus2\Sigma_4$ $2(1)\quad2(1)\quad2(1)$	$2\Sigma_1\oplus2\Sigma_3\oplus2\Sigma_4$ $2(1)\quad2(1)\quad2(1)$
S	$2S_1\oplus2S_2\oplus2S_4$ $2(1)\quad2(1)\quad2(1)$	$2S_1\oplus2S_3\oplus2S_4$ $2(1)\quad2(1)\quad2(1)$
Z	$2Z_1\oplus2Z_2\oplus2Z_4$ $2(1)\quad2(1)\quad2(1)$	$2Z_1\oplus2Z_3\oplus2Z_4$ $2(1)\quad2(1)\quad2(1)$
Q	$3Q_1\oplus3Q_2$ $3(1)\quad3(1)$	$3Q^+\oplus3Q^-$ $3(1)\quad3(1)$

Note.   The labels given in column 1 for the special points and lines of symmetry in the Brillouin zone can be identified from fig. 8.   The labels in column 2 are based on the labels used for point-group representations by Koster *et al.* (1963) (with Γ replaced by the letter for the point or line of symmetry in question) ; the labels in column 3 are those used by Bouckaert *et al.* (1936).

This equation can be used for each of the atoms in one fundamental unit cell, that is for each κ and for a fixed l, to determine the matrix that represents the effect of $\{R|\mathbf{v}\}$ on the vector $\boldsymbol{\xi}^{\mathbf{k}}$ comprising the displacements of all the atoms in the unit cell when the crystal is vibrating in such a manner as to involve normal modes with only one wave vector $\mathbf{k}$. For our purposes it is adequate to find the character of this matrix and it is only the diagonal elements of the matrix which contribute to the character. Therefore, eqn. (6.36) means that if an atom on site κ is moved to a different site κ', whether in the same unit cell or not, the contribution of that atom to the character of $\{R|\mathbf{v}\}$ in $\Gamma_{\xi}^{\mathbf{k}}$ will be zero. If eqn. (6.36) is written in terms of components we have

$$\sum_{\beta} R_{\alpha\beta} u_{\beta}^{\mathbf{k}}(l\kappa) = \exp\left[i\mathbf{k}\,.\,\{\mathbf{r}(l'\kappa') - \mathbf{r}(l\kappa')\}\right]u_{\alpha}'^{\mathbf{k}}(l'\kappa') \tag{6.37}$$

so that the character of $\{R|\mathbf{v}\}$ in $\Gamma_{\xi}^{\mathbf{k}}$ is given by

$$\chi_{\xi}^{\mathbf{k}}(\{R|\mathbf{v}\}) = \sum_{\alpha}\sum_{\beta}\sum_{\kappa} \delta_{\alpha\beta}\delta_{\kappa\kappa'} \exp\left[i\mathbf{k}\,.\,\{\mathbf{r}(l\kappa) - \mathbf{r}(l'\kappa)\}\right]R_{\alpha\beta} \tag{6.38}$$

where $\delta_{\kappa\kappa'}$ is used to ensure that the summation over κ is restricted to those atoms which remain on the same site κ, but possibly in different unit cells, under the action of $\{R|\mathbf{v}\}$. Having found the characters of $\Gamma_{\xi}^{\mathbf{k}}$ for any given wave vector $\mathbf{k}$ in the Brillouin zone the symmetries of the normal modes of vibration with wave vector $\mathbf{k}$ can be determined from the reduction of $\Gamma_{\xi}^{\mathbf{k}}$.

As examples of the application of the above procedure we quote in tables 25 and 26 the symmetry assignments of the normal modes of vibration of the

Table 26.   Phonon symmetries in FeF_2

Point	Phonons ($P4_2/mnm$)
Γ	$\Gamma_1^+\oplus\Gamma_2^+\oplus\Gamma_3^+\oplus\Gamma_4^+\oplus\Gamma_5^+\oplus2\Gamma_2^-\oplus2\Gamma_3^-\oplus4\Gamma_5^-$
M	$M_1^+\oplus M_2^+\oplus M_3^+\oplus M_4^+\oplus M_5^+\oplus2M_3^-\oplus2M_4^-\oplus4M_5^-$
Z	$3Z_1\oplus Z_2\oplus3Z_3\oplus2Z_4$
A	$3A_1\oplus A_2\oplus3A_3\oplus2A_4$
R	$3R_1^+\oplus6R_1^-$
X	$6X_1\oplus3X_2$
Δ	$6\Delta_1\oplus6\Delta_2\oplus3\Delta_3\oplus3\Delta_4$
U	$9U_1$
Λ	$3\Lambda_1\oplus\Lambda_2\oplus\Lambda_3\oplus3\Lambda_4\oplus5\Lambda_5$
V	$3V_1\oplus3V_2\oplus V_3\oplus V_4\oplus5V_5$
Σ	$6\Sigma_1\oplus6\Sigma_2\oplus\Sigma_3\oplus5\Sigma_4$
S	$6S_1\oplus3S_2\oplus4S_3\oplus5S_4$
Y	$6Y_1\oplus6Y_2\oplus3Y_3\oplus3Y_4$
T	$9T_1$
W	$9W_1$

Note.   The phonon symmetries given by Katiyar (1970) have been translated into the notation of Dimmock and Wheeler (1962 a). It should be noted that the following pairs of representations become degenerate in the grey space group $P4_2/mnm1'$:

M_1^+, M_2^+ ; M_3^+, M_4^+ ; M_3^-, M_4^- ; V_1, V_2 ; V_3, V_4 ; Y_1, Y_2 ; Y_3, Y_4.

sodium chloride structure (Cochran *et al.* 1966, Sakurai *et al.* 1968) and the rutile structure (Katiyar 1970), while similar identifications for some other common structures will be found in the literature (for example, the perovskite structure (Cowley 1964), the fluorite structure (Dolling *et al.* 1965), the CsCl structure (Warren 1968), the zincblende and diamond structures (Montgomery 1969) and KH_2PO_4 (Montgomery and Paul 1971).

Having determined the irreducible representations to which the normal modes with wave vector $\mathbf{k}$ belong, it is then possible to use the fact that the normal-mode eigenvector, or polarization vector, $e_\alpha(\kappa|\mathbf{k}j)$ belongs to a certain space-group irreducible representation $\Gamma_p^{\mathbf{k}}$ to determine $e_\alpha(\kappa|\mathbf{k}j)$. Suppose that $e_\alpha(\kappa|\mathbf{k}j)$ forms one component of a basis of the representation $\Gamma_p^{\mathbf{k}}$ of $\mathbf{G}$ and that we denote this basis by $|u_{p1}^{\mathbf{k}}, u_{p2}^{\mathbf{k}}, ..., u_{ph}^{\mathbf{k}}\rangle$, or $|u_{pi}^{\mathbf{k}}\rangle$ $(i = 1, 2, ..., h$ where h is the dimension of $\Gamma_p^{\mathbf{k}}$). One can use the transformation properties of this basis, that is

$$\{R|\mathbf{v}\}u_{pi}^{\mathbf{k}} = \sum_j u_{pj}^{\mathbf{k}}\mathbf{D}_p^{\mathbf{k}}(\{R|\mathbf{v}\})_{ji} \qquad (6.39)$$

to determine functions $u_{pi}^{\mathbf{k}}$ each of which is a vector of dimension $3r$, where $\mathbf{D}_p^{\mathbf{k}}(\{R|\mathbf{v}\})$ is the matrix representative of $\{R|\mathbf{v}\}$ in $\Gamma_p^{\mathbf{k}}$. The left-hand side of this equation can be evaluated using the known properties of the operator $\{R|\mathbf{v}\}$ (see eqns. (6.26)–(6.29)). Since the matrix representatives $\mathbf{D}_p^{\mathbf{k}}(\{R|\mathbf{v}\})$ are available from the various tables of space-group representations, it is possible to use eqn. (6.39) for many different operators $\{R|\mathbf{v}\}$ to determine $|u_{pi}^{\mathbf{k}}\rangle$ for any given p and $\mathbf{k}$. Equation (6.39) only enables relations between the displacements of atoms of the same chemical element to be determined. Relations between the displacements of atoms of different chemical elements have to be determined from the use of the eigenvalue equation. Instead of using eqn. (6.39) directly, the determination of bases $|u_{pi}^{\mathbf{k}}\rangle$ of $\Gamma_p^{\mathbf{k}}$ can be performed using projection operator techniques, see for example, Montgomery (1969). If $\Gamma_p^{\mathbf{k}}$ occurs only once among the representations obtained in the reduction of $\Gamma_\xi^{\mathbf{k}}$, $e_\alpha(\kappa|\mathbf{k}j)$ can immediately be taken to be $|u_{pi}^{\mathbf{k}}\rangle$. However, if $\Gamma_p^{\mathbf{k}}$ occurs c_p times in the reduction of $\Gamma_\xi^{\mathbf{k}}$, where $c_p > 1$, the normal mode eigenvectors $e_\alpha(\kappa|\mathbf{k}j)$ will be given by some linear combination of the c_p bases $|^1u_{pi}^{\mathbf{k}}\rangle$, $|^2u_{pi}^{\mathbf{k}}\rangle$, ..., $|^{c_p}u_{pi}^{\mathbf{k}}\rangle$, which have been found to satisfy eqn. (6.39). That is, we write

$$e_\alpha^i(\kappa|\mathbf{k}j) = \sum_{\lambda=1}^{c_p} {}^\lambda b_p^{\mathbf{k}}\,{}^\lambda u_{pi}^{\mathbf{k}} \qquad (6.40)$$

where $e_\alpha^i(\kappa|\mathbf{k}j)$ belongs to row i of $\Gamma_p^{\mathbf{k}}$. The coefficients ${}^\lambda b_p^{\mathbf{k}}$ cannot be found group-theoretically ; they depend on the force constants between the various atoms in the crystal and therefore have to be found from the diagonalization of the dynamical matrix $D_{\alpha\beta}(l\kappa ; l'\kappa')$. Examples of the determination of the eigenvectors $e_\alpha(\kappa|\mathbf{k}j)$ have been given in considerable detail by Cowley (1964), Maradudin and Vosko (1968), Warren (1968), Montgomery (1969), and Katiyar (1970) for some simple materials. The purpose behind the group-theoretical determination of the eigenvectors is that, as for molecules, it enables one to identify the form of the vibrations in any given normal mode. Also, in those cases in which it is meaningful to describe modes as longitudinal or transverse, it is possible to identify the various modes as longitudinal or transverse by inspection of the eigenvectors.

The above discussion has been given in terms of the displacements of the atoms and the use of classical physics in the form of Newton's laws of motion. However, the discussion could also be given in terms of quantum mechanics and the most convenient way of doing this is to make use of the creation and annihilation operators which are commonly used in connection with the quantum-mechanical treatment of a harmonic oscillator. For a single one-dimensional harmonic oscillator with operators x and p for the position and momentum coordinates it is often convenient to use the boson creation and annihilation operators $a^\dagger$ and a defined by

$$\left.\begin{aligned}a^\dagger &= (2m\hbar\omega)^{-1/2}(p + im\omega x)\\[2mm]a &= (2m\hbar\omega)^{-1/2}(p - im\omega x)\end{aligned}\right\} \tag{6.41}$$

where m is the mass of the particle and $\omega/2\pi$ is the frequency of the oscillations. We have previously expanded the displacement $u_\alpha(l\kappa)$ in terms of the displacements in the normal modes

$$u_\alpha(l\kappa) = (M_\kappa)^{-1/2} \sum_j \sum_{\mathbf{k}} e_\alpha(\kappa|\mathbf{k}j)Q_j(\mathbf{k}) \exp\left[i\{\mathbf{k}\cdot\mathbf{r}(l\kappa) - \omega_j(\mathbf{k})t\}\right] \tag{6.33}$$

and one can introduce a similar expansion for the momentum $p_\alpha(l\kappa)$

$$p_\alpha(l\kappa) = (M_\kappa)^{1/2} \sum_j \sum_{\mathbf{k}} p_\alpha(\kappa|\mathbf{k}j)P_j(\mathbf{k}) \exp\left[i\{\mathbf{k}\cdot\mathbf{r}(l\kappa) - \omega_j(\mathbf{k})t\}\right]. \tag{6.42}$$

After some manipulation the Hamiltonian for the vibrations of the atoms in the crystal, in the harmonic approximation, can be written in the form

$$\mathscr{H} = \tfrac{1}{2} \sum_j \sum_{\mathbf{k}} \{P_j(\mathbf{k})P_j(-\mathbf{k}) + \omega_j(\mathbf{k})^2 Q_j(\mathbf{k})Q_j(-\mathbf{k})\}. \tag{6.43}$$

This equation gives the Hamiltonian $\mathscr{H}$ in terms of the normal coordinates of the system and it can be regarded as the Hamiltonian of a system of non-interacting, or de-coupled, harmonic oscillators, with each oscillator characterized by a particular set of values of j and $\mathbf{k}$. We can define a pair of creation and annihilation operators $a_{j\mathbf{k}}{}^\dagger$ and $a_{j\mathbf{k}}$ for each normal mode by

$$\left.\begin{aligned}a_{j\mathbf{k}}{}^\dagger &= \{2\hbar\omega_j(\mathbf{k})\}^{-1/2}\{\omega_j(\mathbf{k})Q_j(-\mathbf{k}) - iP_j(\mathbf{k})\}\\[2mm]a_{j\mathbf{k}} &= \{2\hbar\omega_j(\mathbf{k})\}^{-1/2}\{\omega_j(\mathbf{k})Q_j(\mathbf{k}) + iP_j(-\mathbf{k})\}\end{aligned}\right\} \tag{6.44}$$

which are similar to the operators $a^\dagger$ and a defined for a single harmonic oscillator in eqn. (6.41). The symmetry of the normal mode specified by j, $\mathbf{k}$ could then be determined by studying the transformation properties of $a_{j\mathbf{k}}{}^\dagger$ and $a_{j\mathbf{k}}$ under the operations $\{R|\mathbf{v}\}$ of the space group $\mathbf{G}$ of the crystal and thereby assigning these operators to the appropriate space-group irreducible representations $\Gamma_p{}^{\mathbf{k}}$. We mention this procedure here not so much because this method is actually used in connection with phonon symmetries, but rather because it provides a natural and convenient introduction to the study of magnon symmetries, which we shall discuss in the next section and for which the quantum-mechanical treatment is rather more convenient than a classical approach.

So far in this section we have not mentioned the operation of time-reversal. For a crystalline material which does not exhibit spontaneous magnetic ordering it can be assumed that θ, the operation of time-reversal, is a symmetry operation of a specimen of the material. The group **G** that would be used in connection with labelling the branches of the phonon dispersion relations or determining the eigenvectors for the various normal modes would then be one of the 230 grey space groups. For a crystal which does exhibit magnetic ordering there may be some choice as to the group to choose as **G** in this group-theoretical analysis. If it is assumed that the lattice vibrations can be studied independently of any changes in the orientations of the magnetic moments in the specimen, **G** will be the space group of the atomic positions in the crystal. However, if the interactions between the lattice vibrations and the motions of the spins are deemed to be important, it may be more appropriate to use the magnetic space group of the crystal as **G** in the group-theoretical analysis of the normal modes of vibration (see § 6.8).

6.4. *Magnon symmetries*†

It is possible to extend to magnons the classification into acoustic and optic modes which is commonly used for phonons and which we have mentioned at the beginning of the previous section. For any crystal there must be three lattice vibrational modes which can be described approximately in terms of the macroscopic elastic constants of the crystal for long wavelength, that is for small **k**. The dispersion relation for each of the acoustic modes will take the form of a straight line passing through the origin. If we adopt the rule that all the remaining normal modes for any given crystal are described as optic modes it can be seen that the classification into acoustic and optic modes for phonons is of general validity. Whereas the existence of the acoustic modes is a general feature of the phonon dispersion relations for any crystal, a casual inspection of a selection of published magnon dispersion relations reveals that the existence of modes with the characteristic ' acoustic-mode ' properties, namely with the simple linear dispersion relation $\omega_j(\mathbf{k}) = C_1|\mathbf{k}|$ in the vicinity of $\mathbf{k} = \mathbf{0}$ where $\omega_j(\mathbf{k})/2\pi$ is the magnon frequency and C_1 is a constant, is by no means universal. In particular the dependence of $\omega_j(\mathbf{k})$ on $|\mathbf{k}|$ is very often not linear and there may also be a gap, that is $\omega_j(\mathbf{k}) \neq \mathbf{0}$ at $\mathbf{k} = 0$. The gap is often sufficiently large that the energies of the magnons in the vicinity of $\mathbf{k} = \mathbf{0}$ correspond to infra-red frequencies, or very high ultrasonic frequencies, rather than to ordinary acoustic frequencies. Nevertheless, in spite of the reduced physical significance of the distinction, it is quite common to adopt for magnons a rule that is similar to that used

---

† *Late extra* : Professor Coles has brought to my notice before publication an article ' On the application of group theory to spin waves in collinear magnetic structures ' by V. C. Sahni and G. Venkataraman appearing in *Advances in Physics* (1974) **23,** 547. This paper gives a much more formal and detailed treatment of the subject of spin-wave symmetries than is contained in the present section. The article by Sahni and Venkataraman really is the analogue for spin waves of the comprehensive work done for phonons by Maradudin and Vosko (1968) to which we referred in § 6.3.

for phonons ; that is, we describe the lowest spin-wave mode at each **k** as an acoustic magnon mode (MA) and all the remaining modes as optic magnon modes (MO). Since we are dealing with only one degree of freedom the number of acoustic magnon branches to the dispersion relations for any given crystal is 1 and not **3**.

A more versatile classification scheme for magnon symmetries can be developed by using group theory. The same general theorem, namely Wigner's theorem, which lies behind the use of group theory in labelling the electronic band structure and the phonon dispersion relations, also applies to the magnon dispersion relations. That is to say, if $\mathscr{H}$ is the Hamiltonian used to describe a magnetically-ordered crystal, any given magnon must belong to one or other of the irreducible representations, or co-representations, of the group of $\mathscr{H}$, which in turn will be determined by the symmetry of the crystal itself.

In discussing the application of group theory to the labelling of spin-wave dispersion relations and to the study of essential degeneracies in these dispersion relations, it is convenient to use a quantum-mechanical approach from the outset and not to proceed via a classical, or semi-classical, treatment. For a magnetically ordered material it is common to separate from the total Hamiltonian of the material that part $\mathscr{H}_S$, usually called a ' spin Hamiltonian ', which includes all the terms which involve the magnetic moments in the system and no other terms. The frequencies of the spin waves, which are collective excitations of the magnetic moments in the system, can be determined by finding the eigenvalues of the spin Hamiltonian. One of the simplest forms of spin Hamiltonian $\mathscr{H}_S$ that is encountered is the Heisenberg Hamiltonian which takes the form

$$\mathscr{H}_S = -J_{ij} \sum_{i \neq j} \mathbf{S}_i \cdot \mathbf{S}_j \qquad (6.45)$$

for a simple ferromagnet with only one magnetic constituent.

As a first step towards diagonalizing a spin Hamiltonian it is common to make use of the Holstein–Primakoff transformation in which we write the spin operators S_i^+ ($=S_{ix}+iS_{iy}$) and S_i^- ($=S_{ix}-iS_{iy}$) in terms of boson creation and annihilation operators $a_i^\dagger$ and a_i, which are similar to the operators used in connection with lattice vibrations (Holstein and Primakoff 1940)

$$\left.\begin{aligned} S_i^+ &= (2S)^{-1/2}(1 - a_i^\dagger a_i/2S)^{1/2}a_i \\ S_i^- &= (2S)^{-1/2}a_i^\dagger(1 - a_i^\dagger a_i/2S)^{1/2} \end{aligned}\right\}. \qquad (6.46)$$

From these one can obtain explicit expressions for S_{ix} and S_{iy} directly while after some manipulation, using the fact that $S_{ix}^2 + S_{iy}^2 + S_{iz}^2 = S(S+1)$ together with certain commutation relations of the a_i and $a_i^\dagger$ operators, S_{iz} can be obtained as well. Therefore, a spin Hamiltonian $\mathscr{H}_S$ can be expressed in terms of $a_i^\dagger$ and a_i instead of S_i. If the material is crystalline we can make use of Bloch's theorem and make a further transformation to a set of wave-like operators

$$\left.\begin{aligned} b_{\mathbf{k}}^\dagger &= (N)^{-1/2} \sum_i \exp{(i\mathbf{k} \cdot \mathbf{r}_i)}a_i^\dagger \\ b_{\mathbf{k}} &= (N)^{-1/2} \sum_i \exp{(-i\mathbf{k} \cdot \mathbf{r}_i)}a_i \end{aligned}\right\}. \qquad (6.47)$$

In the linear approximation we retain only the first term in the binomial expansion for each of the square roots on the right-hand side of eqn. (6.46) so that we have

$$S_i{}^+ \fallingdotseq (2S)^{-1/2} a_i \atop S_i{}^- \fallingdotseq (2S)^{-1/2} a_i{}^\dagger \left.\right\}$$ (6.48)

and therefore

$$b_{\mathbf{k}}{}^\dagger \fallingdotseq (2S/N)^{1/2} \sum_i \exp(i\mathbf{k} \cdot \mathbf{r}_i) S_i{}^- \atop b_{\mathbf{k}} \fallingdotseq (2S/N)^{1/2} \sum_i \exp(-i\mathbf{k} \cdot \mathbf{r}_i) S_i{}^+ \left.\right\}.$$ (6.49)

The wave-like operators $b_{\mathbf{k}}{}^\dagger$ and $b_{\mathbf{k}}$ are creation and annihilation operators for magnons with wave vector $\mathbf{k}$ and so the symmetries of the magnons can most conveniently be determined by studying the transformation properties of $b_{\mathbf{k}}{}^\dagger$ and $b_{\mathbf{k}}$. By applying the various space-group operations $\{R|\mathbf{v}\}$ to $b_{\mathbf{k}}{}^\dagger$ and $b_{\mathbf{k}}$ it is possible to assign these operators to one of the irreducible co-representations of the Heesch–Shubnikov space group of the material.

The form of $b_{\mathbf{k}}{}^\dagger$ and $b_{\mathbf{k}}$ in eqn. (6.49) is for a simple ferromagnetic material in which it is assumed that all the spins are on a single sublattice which possesses the full translational symmetry of the Bravais lattice of the crystal. That is, it is assumed that there is only one magnetic atom in each unit cell of the crystal ; this assumption is clearly valid for f.c.c. and b.c.c. ferromagnetic metals and the symmetries of the magnons for all special wave vectors $\mathbf{k}$ throughout the Brillouin zone have been identified and tabulated in both these cases by Cracknell (1970 b). For a hexagonal close-packed (h.c.p.) metal there are two atoms per unit cell and therefore in constructing magnon creation and annihilation operators one must use two sublattices and there will be two, possibly degenerate, spin-wave frequencies for each wave vector, $\mathbf{k}$, in the Brillouin zone. If S_i and S_j are the spin vectors and $\mathbf{r}_i$ and $\mathbf{r}_j$ are the position vectors of the atoms on the two sublattices the two creation operators are

$$b_{1\mathbf{k}}{}^\dagger = (2NS)^{-1/2} \sum_i \exp(i\mathbf{k} \cdot \mathbf{r}_i) S_i{}^- \atop b_{2\mathbf{k}}{}^\dagger = (2NS)^{-1/2} \sum_j \exp(i\mathbf{k} \cdot \mathbf{r}_j) S_j{}^- \left.\right\}.$$ (6.50)

Again, by applying the space-group symmetry operations to these creation operators it is possible to determine the symmetries of the magnons in both branches of the spin wave dispersion relations for all wave vectors $\mathbf{k}$ in the Brillouin zone. The results have been tabulated by Cracknell (1970 b) for the three most important possible directions of the magnetization and it was found that it is only at very few points in the Brillouin zone that the two spin-wave branches are required to be degenerate as a result of symmetry. For a simple two-sublattice antiferromagnetic crystal there will also be distinct creation operators like those in eqn. (6.50) for each sublattice ; in this case the two sublattices are distinguished by involving atoms which are not only on different crystallographic sites, but also have spins in opposite orientations in the ground state. The results for antiferromagnetic MnF_2

were deduced by Loudon (1968) and are reproduced in table 27, where the notation is based on that of Dimmock and Wheeler (1962 a) for this material and the unit cell and Brillouin zone for this material are illustrated in figs. 1 and 6(b). The magnon symmetries in one or two other simple antiferromagnetic crystals have also been determined (NiF_2, Joshua and Cracknell (1969), CoO and NiO, Daniel and Cracknell (1969), Cracknell and Joshua (1969 a), UO_2, Cracknell and Joshua (1969 b)).

For each of the ferromagnetic and simple antiferromagnetic crystals which have been mentioned so far in this section, the space group which has been used in studying the magnon symmetries has always been one of the Heesch–Shubnikov groups. However, this need not always be the case and other generalized symmetry groups (see § 1.2) could be used instead if they were considered to be more appropriate for use in the description of the symmetry of the magnetically ordered crystal. Also it is possible to study the transformation properties of the operators in eqn. (6.49) or eqn. (6.50) under the

Table 27.   Magnon symmetries in antiferromagnetic MnF_2 or FeF_2

Point	Magnons	
Γ	$\Gamma_3{}^+ + \Gamma_4{}^+$	$D\Gamma_{3,4}{}^+$
M	$M_3{}^+,\ M_4{}^+$	$DM_3{}^+,\ DM_4{}^+$
Z	Z_2	DZ_2
A	A_2	DA_2
R	$R_1{}^+$	$DR_1{}^+$
X	X_2	DX_2
Δ	$\Delta_3,\ \Delta_4$	$D\Delta_3,\ D\Delta_4$
U	U_1	DU_1
Λ	$\Lambda_3 + \Lambda_4$	$D\Lambda_{3,4}$
V	$V_3,\ V_4$	$DV_3,\ DV_4$
Σ	$2\Sigma_2$	$2D\Sigma_2$
S	$S_1,\ S_2$	$DS_1,\ DS_2$
Y	$Y_3,\ Y_4$	$DY_3,\ DY_4$
T	T_1	DT_1
W	W_1	DW_1

Note. In column 2 the magnon symmetries are identified in terms of the irreducible representations of $Pnnm$ (Dimmock and Wheeler 1962 a, Loudon 1968) and in column 3 they are identified in terms of the irreducible co-representations of $P4_2{}'/mnm'$ (see § 7.7 of Bradley and Cracknell (1972)).

symmetry operations of the space group of the non-magnetic phase to study the symmetries of magnetic excitations (' paramagnons ') that may occur in the paramagnetic phase which exists at temperatures higher than the critical temperature for the existence of the magnetically ordered phase of the material (Cracknell 1969 d). In this section we have used the Holstein–Primakoff transformation (eqn. (6.46)) to convert a spin Hamiltonian into a form involving operators that resemble the operators in the Hamiltonian of a lattice vibrational problem. It is also possible to use this transformation for the opposite purpose, that is to transform a lattice vibrational Hamiltonian into a form involving ' pseudo-spin ' operators so that one can then use some of the techniques which are commonly employed in diagonalizing spin Hamiltonians (see, for example, Tokunaga and Matsubara (1966) and Montgomery and Paul (1971) on the vibrations of the H atoms in KH_2PO_4, and various papers on some rare earth vanadates (Elliott *et al.* 1971, Elliott *et al.* 1972, Sandercock *et al.* 1972)).

6.5. *Density of states and other Brillouin-zone integrations*

The arguments that were given in § 6.2 concerning the symmetry of $E_j(\mathbf{k})$ were phrased in terms of the energies of itinerant electrons in a solid. However, the symmetry arguments apply just as well if $E_j(\mathbf{k})$ is the energy of a phonon or magnon instead of the energy of an electron. This symmetry of $E_j(\mathbf{k})$ can be exploited in calculations of the density of states and of other quantities, such as the lattice vibrational specific heat, which involve integrating the contributions from states with all possible wave vectors in the Brillouin zone.

The problem of the determination of some quantity involving contributions from states with all wave vectors in the Brillouin zone involves evaluating an integral of the form

$$F = \sum_j \int \int \int G\{E_j(\mathbf{k})\}k^2 \sin\theta \, dk \, d\theta \, d\phi. \qquad (6.51)$$

We take $E_j(\mathbf{k})$ to be the energy of an electron or of a phonon or of a magnon. The form of the function $G\{E_j(\mathbf{k})\}$ will vary, depending on the property under investigation ; for the density of states it will involve the delta function $\delta(E - E_j(\mathbf{k}))$. For the lattice-vibrational specific heat $G\{E_j(\mathbf{k})\}$ will involve $E_j(\mathbf{k})$, the energy of a phonon with wave vector $\mathbf{k}$ on branch j of the dispersion relations, and the derivative with respect to temperature of the Bose–Einstein distribution function. $\mathbf{k}$ may be written in spherical polar coordinates as (k, θ, ϕ) and we can re-arrange eqn. (6.51) and, formally, integrate over k ; thus

$$F = \sum_j \int \int [\int G\{E_j(\mathbf{k})\}k^2 \, dk] \sin\theta \, d\theta \, d\phi$$

$$= \sum_j \int \int G_j(\theta, \phi) \sin\theta \, d\theta \, d\phi \qquad (6.52)$$

where

$$G_j(\theta, \phi) = \int G\{E_j(\mathbf{k})\}k^2 \, dk. \qquad (6.53)$$

$G_j(\theta, \phi)$ represents the line integral of $G\{E_j(\mathbf{k})\}$ in the direction (θ, ϕ) from Γ to the Brillouin-zone boundary.

One could parametrize $E_j(\mathbf{k})$ by the method suggested in § 6.2 and then use the resultant expression to calculate $E_j(\mathbf{k})$ for all $\mathbf{k}$ and hence evaluate F using eqn. (6.51). In practice, however, the data which may be available for $E_j(\mathbf{k})$ may be restricted to the values of $E_j(\mathbf{k})$ along certain special lines of symmetry in the Brillouin zone. For example, this is frequently the case in the results of the determination of phonon energies from inelastic neutron-scattering experiments. It would be rather optimistic to hope to be able to determine reliable values of the coefficients in a geometrical parametrization of $E_j(\mathbf{k})$ from such restricted data. It is possible to avoid this difficulty, to some extent at least, by adopting an alternative procedure which was introduced for the lattice vibrational specific heat by Houston (1948) and which has subsequently been developed by a number of other authors.   Houston's method has the advantage that it can be used to give an expression for F in eqn. (6.51) in terms of line integrals along just those lines of symmetry along which the energies $E_j(\mathbf{k})$ are available.   The method involves some exploitation of the symmetry of $E_j(\mathbf{k})$ as a function of $\mathbf{k}$. $E_j(\mathbf{k})$ can be regarded as a function that is centred at Γ and possesses the symmetry of the point group $\mathbf{P}'$ (see § 6.2).   Therefore $G\{E_j(\mathbf{k})\}$ and $G_j(\theta, \phi)$ are also functions which possess the symmetry of the point group $\mathbf{P}'$.   Each of these functions can, therefore, be expanded in terms of functions which possess the symmetry of $\mathbf{P}'$; in particular we expand $G_j(\theta, \phi)$ so that

$$G_j(\theta, \phi) = \sum_i C_i{}^j \chi_i(\theta, \phi) \tag{6.54}$$

where the functions $\chi_i(\theta, \phi)$ are orthogonal to one another.   Houston considered the example of a simple cubic monatomic structure.   In this case $\chi_i(\theta, \phi)$ will be the functions listed in table 12 as belonging to the identity representation ($\Gamma_1{}^+$ or A_{1g}) of $m3m$.   The argument then proceeds by replacing the integral in eqn. (6.52), over all directions (θ, ϕ), by a summation of the line integrals $G_j(\theta, \phi)$ over those particular directions corresponding to special lines of symmetry for which $E_j(\mathbf{k})$, and therefore also $G\{E_j(\mathbf{k})\}$, is actually known.   That is, we write

$$F = \sum_j \sum_\lambda C_{j\lambda} G_j(\theta_\lambda, \phi_\lambda) \tag{6.55}$$

where λ labels the lines of symmetry.   Using eqn. (6.54) this becomes

$$F = \sum_j \sum_\lambda C_{j\lambda}\{ \sum_i C_i{}^j \chi_i(\theta_\lambda, \phi_\lambda)\}$$
$$= \sum_j \sum_i \{ \sum_\lambda C_{j\lambda} \chi_i(\theta_\lambda, \phi_\lambda)\} C_i{}^j. \tag{6.56}$$

But using eqns. (6.52) and (6.54) we also have

$$F = \sum_j \int\int \{ \sum_i C_i{}^j \chi_i(\theta, \phi)\} \sin\theta \, d\theta \, d\phi$$
$$= \sum_j 4\pi C_1{}^j \tag{6.57}$$

where we have taken $\chi_1(\theta, \phi)$ to be $P_0{}^0(\cos\theta)$ and used the orthogonality of the other $\chi_i(\theta, \phi)$ to $P_0{}^0(\cos\theta)$.   Equating coefficients of $C_i{}^j$ in eqns. (6.56) and (6.57) we obtain

$$\left.\begin{aligned} \sum_\lambda C_{j\lambda} &= 1 \\ \sum_\lambda C_{j\lambda} \chi_i(\theta_\lambda, \phi_\lambda) &= 0 \quad (i \neq 1) \end{aligned}\right\}. \tag{6.58}$$

The eqns. (6.58) can be solved to give the values of the constants $C_{j\lambda}$.   These values of $C_{j\lambda}$ can then be substituted into eqn. (6.55) to give the value of the required quantity F.

Applications of Houston's method have generally been concerned with cubic crystals and with the phonon contribution to the specific heat. However, the method has been extended to some tetragonal crystals by Begum *et al.* (1969) who used it for calculating the spin-wave contributions to the specific heats of the antiferromagnetic difluorides of some transition metals.

6.6. *Van Hove singularities and zero-slope points*

The existence of singularities, or more correctly of cusps, in the density of states, $n(E)$, was first studied in connection with the lattice vibrations of a crystalline solid, where $E_j(\mathbf{k})$ is the energy of a phonon with wave vector $\mathbf{k}$. For a two-dimensional crystal the density of states $n(E)$ for the lattice vibrations possesses logarithmic singularities; for a three-dimensional problem $n(E)$ remains finite but its first derivative exhibits infinite discontinuities (Van Hove 1953). Although these singularities, which are now commonly referred to as 'Van Hove singularities', were first noticed in connection with the phonon density of states, their existence is not confined to the phonon density of states but they should be expected to occur in the density of states for any particle, or quasiparticle excitation, in a crystalline solid.

The peaks in the density of states occur at those frequencies at which $E_j(\mathbf{k})$, as a function of the three coordinates k_x, k_y and k_z, is stationary; this is because at such points in $\mathbf{k}$ space there are a large number of allowed states with nearly identical values of the energy. If we use $E_j(\mathbf{k})$ to denote the energy of a particle, or quasiparticle excitation, with wave vector $\mathbf{k}$, its stationary values will occur when

$$\delta E_j(\mathbf{k}) = \frac{\partial E_j(\mathbf{k})}{\partial k_x}\,\delta k_x + \frac{\partial E_j(\mathbf{k})}{\partial k_y}\,\delta k_y + \frac{\partial E_j(\mathbf{k})}{\partial k_z}\,\delta k_z = 0 \qquad (6.59)$$

that is when

$$\partial E_j(\mathbf{k})/\partial k_x = \partial E_j(\mathbf{k})/\partial k_y = \partial E_j(\mathbf{k})/\partial k_z = 0. \qquad (6.60)$$

Values of $\mathbf{k}$ at which $E_j(\mathbf{k})$ is stationary are sometimes called 'zero-slope' points (Kudryavtseva 1967). There are two different kinds of stationary values of $E_j(\mathbf{k})$ which we may describe as 'accidental' and 'essential' stationary values. An accidental stationary value will normally occur at a general, i.e. arbitrary, wave vector with no special symmetry. If a small change is made in the potential term in the Hamiltonian of the system, the location of such a stationary value of $E_j(\mathbf{k})$ will normally be altered. Group-theoretical arguments cannot usefully be applied to these accidental stationary values. An essential stationary value is one which occurs for a certain wave vector as a result of the symmetry of the crystal; this wave vector will correspond to one of the points of symmetry in the Brillouin zone. A small change in the potential may alter the shapes of the $E_j(\mathbf{k})$ functions but these stationary values of $E_j(\mathbf{k})$ will still occur at the same wave vectors as before. Group-theoretical arguments can be used to identify the wave vectors at which essential stationary values may be expected to occur. Both kinds of stationary values of $E_j(\mathbf{k})$ give rise to Van Hove singularities. However, since group-theoretical arguments can only be applied to essential stationary values, we shall not discuss accidental stationary values any further.

In addition to the identification of those wave vectors **k** at which essential stationary values can be expected to occur, there is a slightly different, but nevertheless closely related, problem which we ought to consider. There may be many points **k** in the Brillouin zone for which, although $\delta E_j(\mathbf{k})$ is not equal to zero and therefore there is not a stationary value of $E_j(\mathbf{k})$ there, nevertheless $\partial E_j(\mathbf{k})/\partial(\mathbf{k} \cdot \mathbf{n})$ may be zero for some particular direction specified by the unit vector **n**. This is commonly observed in the fact that, at least for crystals of high symmetry, the electronic energy bands or the phonon dispersion relations become flat as they approach the Brillouin-zone boundary in some principal direction. Indeed, because this happens for all the principal directions in a number of common crystals, it has sometimes been supposed that energy bands or phonon dispersion relations always become flat as they approach a Brillouin-zone boundary. This is not so and we shall examine this question in a little more detail. Once again we shall only be concerned with the situation in which $\partial E_j(\mathbf{k})/\partial(\mathbf{k} \cdot \mathbf{n})$ vanishes as a result of symmetry and not with the situation in which it vanishes accidentally.

We have already devoted some attention in § 6.2 to the identification of the symmetry of $E_j(\mathbf{k})$ as a function of **k**. In particular we noted that it may be convenient to consider the point-group symmetry of $E_j(\mathbf{k})$ referred to some point $\mathbf{k}_0$; these point groups $\mathbf{P}'(\mathbf{k}_0)$ have been identified by Cracknell (1973 a, b) for a crystal belonging to any possible classical space group or to any possible magnetic space group. We shall now see how the symmetry properties of $E_j(\mathbf{k})$ can be exploited in searching for stationary values of $E_j(\mathbf{k})$. We are looking for values of $\mathbf{k}_0$ such that when $E_j(\mathbf{k})$ is expanded as a Taylor series about $\mathbf{k}_0$ the first-order terms in the expansion vanish, see eqns. (6.59) and (6.60). In the case of 'essential' stationary values this vanishing occurs as a result of the symmetry of $E_j(\mathbf{k})$ and is independent of the Hamiltonian which is used in calculating the band structure or the phonon dispersion relations. Normally the results of the determination of the band structure of a material are presented by choosing a particular straight line with direction specified by the unit vector **n** so that we can write

$$\mathbf{k} = \mathbf{k}_0 + \kappa\mathbf{n} \tag{6.61}$$

along this line. Curves of $E_j(\mathbf{k})$ against κ are then plotted. These curves are usually drawn for lines which join points of high symmetry in the Brillouin zone. In the vicinity of a given wave vector $\mathbf{k}_0$ the function $E_j(\mathbf{k})$ has the symmetry of $\mathbf{P}'(\mathbf{k}_0)$. We can study the behaviour of the function $E_j(\mathbf{k})$ in the vicinity of $\mathbf{k}_0$ along directions $\mathbf{n}_1$, $\mathbf{n}_2$ and $\mathbf{n}_3$ chosen parallel to the three Cartesian axes k_x, k_y and k_z. If $\mathbf{k}_0$ is an arbitrary wave vector within the Brillouin zone these curves will have no special symmetry about $\mathbf{k}_0$; in other words $\mathbf{P}'(\mathbf{k}_0)$ is just the point group 1 (C_1). In this case there will be no essential stationary value of $E_j(\mathbf{k})$ at $\mathbf{k}_0$.

It is, however, possible that for some direction **n** the energy band $E_j(\mathbf{k})$ may satisfy the condition

$$E_j(\mathbf{k}_0 + \kappa\mathbf{n}) = E_j(\mathbf{k}_0 - \kappa\mathbf{n}) \tag{6.62}$$

and, assuming this does not happen accidentally, this means that when plotted along the line $\mathbf{k}_0 + \kappa\mathbf{n}$ the energy band $E_j(\mathbf{k})$ will be symmetrical

Fig. 16

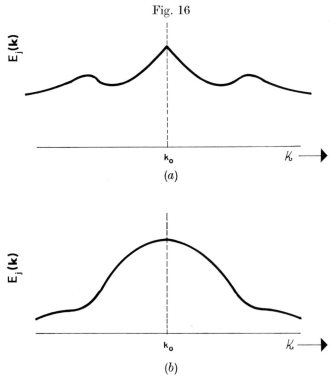

Schematic energy band $E_j(\mathbf{k})$ with $E_j(\mathbf{k}_0+\kappa\mathbf{n})=E_j(\mathbf{k}_0-\kappa\mathbf{n})$.

about $\mathbf{k}_0$ in the vicinity of $\mathbf{k}_0$, see fig. 16. Therefore

$$\left[\frac{\partial E_j(\mathbf{k})}{\partial(\mathbf{k}\cdot\mathbf{n})}\right]_{\mathbf{k}_0-\kappa\mathbf{n}}=-\left[\frac{\partial E_j(\mathbf{k})}{\partial(\mathbf{k}\cdot\mathbf{n})}\right]_{\mathbf{k}_0+\kappa\mathbf{n}} \tag{6.63}$$

and, unless there is some reason to suppose that $\nabla_\mathbf{k} E_j(\mathbf{k})$ is discontinuous at $\mathbf{k}_0$, then

$$\frac{\partial E_j(\mathbf{k})}{\partial(\mathbf{k}\cdot\mathbf{n})}=0 \tag{6.64}$$

at $\mathbf{k}_0$. However, if we allow $\nabla_\mathbf{k} E_j(\mathbf{k})$ to be discontinuous at $\mathbf{k}_0$ we simply have a cusp at $\mathbf{k}_0$. The possibility of a cusp cannot be completely ignored because of the extremely important case in which it occurs, namely at Γ for the acoustic branches of the phonon dispersion relations of any crystalline solid. However, apart from this case, we normally expect eqn. (6.62) to imply the result stated in eqn. (6.64), provided that $E_j(\mathbf{k})$ is not degenerate at $\mathbf{k}_0$. Complications may arise when $E_j(\mathbf{k})$ is degenerate at $\mathbf{k}_0$ and this will be illustrated later by an example.† Assuming that this feature is independent of the choice of the potential $V(\mathbf{r})$ and therefore does not arise accidentally, this means that eqn. (6.62) is associated with the existence of

---

† These complications were not considered by Cracknell (1973 a, b).

some symmetry operation in $\mathbf{P}'(\mathbf{k}_0)$. The symmetry element that must be present in $\mathbf{P}'(\mathbf{k}_0)$ to give rise to the result in eqn. (6.62) is one of the following :

(i) I, the operation of space inversion,
(ii) a two-fold axis of rotation in some direction normal to $\mathbf{n}$ and passing through $\mathbf{k}_0$,
(iii) a reflection plane of symmetry normal to $\mathbf{n}$ and passing through $\mathbf{k}_0$.

If we are to satisfy eqn. (6.62) for each of three non-coplanar directions $\mathbf{n}_1$, $\mathbf{n}_2$ and $\mathbf{n}_3$, and therefore to have an essential stationary value or a cusp, there must be at least one of the symmetry operations (i)–(iii) present for each of these three directions of $\mathbf{n}$. The only ways in which this can arise among the 32 crystallographic point groups are that $\mathbf{P}'(\mathbf{k}_0)$ must possess either the operation of space inversion, or three mutually orthogonal two-fold axes of symmetry, or three coplanar two-fold axes of symmetry mutually inclined at $60°$. We therefore expect eqn. (6.62) to be valid, for all directions of $\mathbf{n}$, if $\mathbf{P}'(\mathbf{k}_0)$ is any one of the following point groups :

$$\bar{1},\ 2/m,\ mmm,\ 4/m,\ 4/mmm,\ \bar{3},\ \bar{3}m,\ 6/m,\ 6/mmm,\ m3,\ m3m$$

or

$$222,\ 422,\ \bar{4}2m,\ 622,\ 23,\ 432,\ \bar{4}3m$$

or

$$32,\ \bar{6}2m.$$

Armed with this list of point groups it is very easy, by inspection of the tables of $\mathbf{P}'(\mathbf{k}_0)$ (Cracknell 1973 a, b) to identify, for non-degenerate $E_j(\mathbf{k}_0)$, those wave vectors $\mathbf{k}_0$ at which we would expect to find essential stationary values or, exceptionally, cusps in $E_j(\mathbf{k})$.

In addition to those values of $\mathbf{k}_0$ at which it is possible to satisfy eqn. (6.62) for three mutually perpendicular directions of $\mathbf{n}$, there may be other wave vectors $\mathbf{k}_0$ at which eqn. (6.62) can be satisfied for certain directions of $\mathbf{n}$ but not for three non-coplanar directions. This means that if the bands $E_j(\mathbf{k})$ are plotted along some directions $\mathbf{n}$ at $\mathbf{k}_0$ they will be flat, or have a cusp, at $\mathbf{k}_0$. By inspection it is possible to show that there are three possibilities ; they are that eqn. (6.62) holds at $\mathbf{k}_0$ for the following directions of $\mathbf{n}$:

(i) $\mathbf{n}$ is in any direction in one specified plane (2, $mm2$, 4, $\bar{4}$, $4mm$, 6, $6mm$),
(ii) $\mathbf{n}$ is along one of three specified coplanar lines which are mutually inclined at $60°$ ($3m$),
(iii) $\mathbf{n}$ is in one specified direction (m, $\bar{6}$).

These various types of behaviour occur in the point groups indicated in brackets ; the directions of $\mathbf{n}$ can be identified by inspection. The satisfaction of eqn. (6.62), which for non-degenerate $E_j(\mathbf{k}_0)$ implies the vanishing of $\partial E_j(\mathbf{k})/\partial(\mathbf{k}\,.\,\mathbf{n})$, for these restricted directions of $\mathbf{n}$, may not lead to saddle points, or essential stationary values, at $\mathbf{k}_0$. If this does not happen, then, in turn, we would not expect this $E_j(\mathbf{k})$ at $\mathbf{k}_0$ to lead to a Van Hove singularity in the density of states. Nevertheless, the vanishing of $\partial E_j(\mathbf{k})/\partial(\mathbf{k}\,.\,\mathbf{n})$ is obviously important if one is concerned with drawing curves showing the energy bands or phonon dispersion relations for various different directions in the Brillouin zone of a given crystal. Finally, there are two point groups for which, if they occur as $\mathbf{P}'(\mathbf{k}_0)$, we would not expect eqn. (6.62) to hold for any direction of $\mathbf{n}$; they are the point group 3 and the trivial point group 1.

A detailed identification of the behaviour of $E_j(\mathbf{k})$ in the vicinity of $\mathbf{k}_0$, in terms of the various possibilities just described, is given in tables which have been published previously both for the classical space groups (Cracknell 1973 a) and for magnetic space groups (Cracknell 1973 b). As an example of the use of these tables to illustrate the behaviour of $E_j(\mathbf{k})$ we consider the case of a crystal with the diamond structure; this structure belongs to the space group $Fd3m$ (O_h^7) and the Brillouin zone is illustrated in fig. 8. Taking the space group $Fd3m$ to be $\mathbf{G}$ we find the corresponding space group $\mathbf{G}'$ is $Im3m$. An inspection of table 28 reveals that $E_j(\mathbf{k})$ has an essential stationary value, or a cusp, at each of the points of symmetry Γ, X, L and W. This means that as a non-degenerate energy band, or a quasiparticle dispersion curve, approaches one of these points from any direction $E_j(\mathbf{k})$ must flatten. The lines and planes of symmetry provide examples of wave vectors $\mathbf{k}_0$ for which $E_j(\mathbf{k})$ is flat, or has a cusp, at $\mathbf{k}_0$ only for certain directions of $\mathbf{n}$. Thus, suppose that Δ_1 is a fixed point $(\alpha, 0, \alpha)$ ($=\alpha\mathbf{g}_1+\alpha\mathbf{g}_3$) on the line Δ ($=\Gamma X$). Then, using table 28, we see that if Δ_1 is approached along any direction normal to ΓX eqn. (6.62) will be satisfied, but if Δ_1 is approached in any other direction there is no reason to expect eqn. (6.62) to be satisfied. Thus, provided $E_j(\mathbf{k}_0)$ is non-degenerate, we should expect $\partial E_j(\mathbf{k})/\partial(\mathbf{k}\cdot\mathbf{n})$ to vanish if Δ_1 is approached from a direction normal to ΓX but not if it is approached from any other direction. Similarly, suppose that Π is an arbitrary point on any one of the planes of symmetry. Then from table 28 we see that a non-degenerate $E_j(\mathbf{k})$ will only be required to be flat at Π if

Table 28.   The identification of $\mathbf{P}'(\mathbf{k}_0)$ for the space group $Fd3m$
$\mathbf{G}'=Im3m$

$\mathbf{k}_0$	$\mathbf{P}'(\mathbf{k}_0)$	V.H.	
Γ	$m3m$	1	
X	$4/mmm$	1	
L	$\bar{3}m$	1	
W	$\bar{4}2m$	1	
$\Delta(\Gamma X)$	$4mm$	$2a$	$[u0w]$
$\Lambda(\Gamma L)$	$3m$	$2b$	$[1\bar{1}0]$, $[\bar{1}01]$, $[0\bar{1}1]$
$\Sigma(\Gamma\Sigma)$	$mm2$	$2a$	$[u\bar{u}w]$
$S(XS)$	$mm2$	$2a$	$[uv\bar{u}]$
$Z(XW)$	$mm2$	$2a$	$[0vw]$
$Q(LW)$	2	$2a$	$[uvv]$
(ΓLK)	m	$2c$	$[1\bar{1}0]$
(ΓKWX)	m	$2c$	$[001]$
(ΓXUL)	m	$2c$	$[\bar{1}01]$
(XUW)	m	$2c$	$[010]$

Note. The data in the last column are concerned with the identification of points at which eqn. (6.62) is satisfied: 1. Equation (6.62) is satisfied for all directions of $\mathbf{n}$. $2a$, $2b$, $2c$. Equation (6.62) is satisfied for certain directions only, according to (i), (ii) and (iii) given in the text. (From Cracknell (1973 a).)

Π is approached along the direction normal to that plane of symmetry and not if Π is approached from any other direction. In particular, if one approaches Π along the line $\Gamma\Pi$ $\partial E_j(\mathbf{k})/\partial(\mathbf{k} \cdot \mathbf{n})$ will not be required to vanish at the Brillouin-zone boundary. Thus we see that whereas the idea seems commonly to be held among many solid-state physicists that energy bands, or quasiparticle dispersion relations, must be flat when they cross the Brillouin-zone boundary this is by no means always true. As a rough general rule we may say that it is more likely to be true for space groups $\mathbf{G}$ of high symmetry than for those of low symmetry. However, even when this flattening at the Brillouin-zone boundary does occur, we still expect it only to be for points of high symmetry on the boundary, or along certain special directions of $\mathbf{n}$. In any given case the problem has to be examined in detail, using the appropriate table similar to table 28 for the crystal in question.

Fig. 17

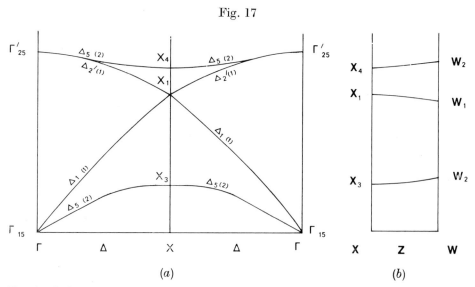

Sketch of phonon dispersion relations for material with diamond structure, (a) along $\Gamma X \Gamma$ for two adjacent Brillouin zones in the repeated zone scheme, (b) along XW (adapted from Montgomery (1969)).

In the above discussion we have assumed that if one can show from symmetry arguments that eqn. (6.62) is satisfied at $\mathbf{k}_0$, then, except for acoustic phonons at Γ, one normally expects to find an essential stationary value of $E_j(\mathbf{k}_0)$ at $\mathbf{k}_0$, provided $E_j(\mathbf{k})$ is non-degenerate at $\mathbf{k}_0$. However, if $E_j(\mathbf{k})$ is degenerate at $\mathbf{k}_0$ the possibility that eqn. (6.62) may imply cusp-like behaviour, rather than stationary values, becomes important. We can illustrate this by considering the example of phonons at the point X in the Brillouin zone of a material with the diamond structure. Since there are two atoms per unit cell in this structure there are six branches to the phonon dispersion relations. The phonon symmetries for this structure have been identified (see Montgomery (1969)) and at X there are three two-fold degenerate branches which are labelled X_1, X_3 and X_4 in the notation of Herring (1942). Suppose we consider the line $\Gamma\Delta X$ and its continuation into the next Brillouin zone, in the repeated zone scheme. The compatibility tables for this space group show that the X_1 branch becomes non-degenerate along Δ, whereas each of the X_3 and X_4 branches remains degenerate along Δ, see fig. 17. From table 28 we see that at X in $Fd3m$ the point group $\mathbf{P}'(\mathbf{k_0})$ is $4/mmm$ and the V.H. behaviour is given as ' 1 '; that is, eqn. (6.62) is satisfied for all directions of $\mathbf{n}$. The two branches which become degenerate at X_1 separate into Δ_1 and Δ_2', which are non-degenerate, as soon as one proceeds away from X along the line Δ. Thus, if one considers just the Δ_1 branch on either side of X it is possible to satisfy eqn. (6.62) by having a cusp at X. Similarly, if one considers just the Δ_2' branch on either side of X it is also possible to satisfy eqn. (6.62) for this branch by having a cusp at X. A single cusp on its own would imply a discontinuity in $\partial E_j(\mathbf{k})/\partial(\mathbf{k}.\mathbf{n})$ which seems to be unlikely in general (except for acoustic phonons at Γ). However, if two branches of $E_j(\mathbf{k})$ are non-degenerate along Δ but become degenerate at X, it is possible to satisfy eqn. (6.62) by having two cusps and yet still avoiding any discontinuity in $\partial E_j(\mathbf{k})/\partial(\mathbf{k}.\mathbf{n})$ at X. This can be achieved by requiring that, if one proceeds along the branch $\Gamma_{15}\Delta_1 X_1$ in one Brillouin zone and then continues along $X_1\Delta_2'\Gamma_{25}'$ in the next Brillouin zone, the slope $\partial E_j(\mathbf{k})/\partial(\mathbf{k}.\mathbf{n})$ remains continuous as one passes from Δ_1 to Δ_2', through X_1. Moreover, if one approaches X_3 or X_4 from any direction except Δ (or Z, along which all the phonon branches must be two-fold degenerate) the two branches which become degenerate at X will be non-degenerate in the vicinity of X. Therefore one would expect that, since eqn. (6.62) must be satisfied for all directions of $\mathbf{n}$ in the vicinity of X, this will again be achieved by having two cusps, but with the values of the slopes fixed so that $\partial E_j(\mathbf{k})/\partial(\mathbf{k}.\mathbf{n})$ remains continuous as one proceeds through X from one branch in the Brillouin zone to the other branch in the next Brillouin zone.

Now let us consider the (acoustic) branch $\Delta_5-X_3-\Delta_5$ and the (optic) branch $\Delta_5-X_4-\Delta_5$, each of which is two-fold degenerate at X and remains two-fold degenerate along Δ. Equation (6.62) still applies and the only way to avoid a discontinuity in $\partial E_j(\mathbf{k})/\partial(\mathbf{k}.\mathbf{n})$ at X is then to suppose that this implies that $\partial E_j(\mathbf{k})/\partial(\mathbf{k}.\mathbf{n})=0$, just as it would if there were any branches that were non-degenerate at $\mathbf{k_0}$. Thus the X_3 and X_4 branches will be flat at X as X is approached along the line Δ. Similarly, since all the phonon branches are two-fold degenerate along Z, all the branches X_1, X_3 and X_4 will be flat at X as X is approached along the line Z. Thus branches X_3 and X_4 have saddle-points at X because for each of them $\partial E_j(\mathbf{k})/\partial(\mathbf{k}.\mathbf{n})=0$ for two orthogonal directions Δ and Z; these two branches therefore can be expected to lead to Van Hove singularities. But for X_1, it was only possible to show that $\partial E_j(\mathbf{k})/\partial(\mathbf{k}.\mathbf{n})=0$ for one direction, namely Z, and this does not imply a saddle-point for X_1; there is therefore no reason to expect X_1 to lead to a Van Hove singularity.

Therefore, summarizing the situation for degenerate $E_j(\mathbf{k})$, suppose that table 28 (or the full tables given by Cracknell (1973 a, b)) show that eqn. (6.62) must be satisfied along a direction $\mathbf{n}$ in the vicinity of $\mathbf{k_0}$. Then if a certain branch of $E_j(\mathbf{k})$ is two-fold degenerate at $\mathbf{k_0}$ and remains degenerate along $\mathbf{n}$ we take eqn. (6.62) to imply that $\partial E_j(\mathbf{k})/\partial(\mathbf{k}.\mathbf{n})=0$ along $\mathbf{n}$, that is, this branch is flat at $\mathbf{k_0}$ as $\mathbf{k_0}$ is approached along $\mathbf{n}$. However, if $E_j(\mathbf{k})$ is two-fold degenerate at $\mathbf{k_0}$ but the degeneracy is lifted along $\mathbf{n}$, then we take eqn. (6.62) to imply the kind of situation illustrated for X_1 in fig. 17. That is, one preserves the continuity of $\partial E_j(\mathbf{k})/\partial(\mathbf{k} \cdot \mathbf{n})$ at $\mathbf{k}_0$, as one passes from one branch to the other through $\mathbf{k}_0$, without requiring the branches to become flat at $\mathbf{k}_0$. These arguments could also be extended for the, rather rare, cases when the degeneracy of $E_j(\mathbf{k})$ at $\mathbf{k}_0$ is greater than two-fold.

The relationship between the approach to Van Hove singularities described in this section and the rather different approach initiated previously by a number of Soviet authors (Karavaev 1964, Kudryavtseva 1967, 1968, 1969, see also Streitwolf (1969 a, b)) has also been examined elsewhere (Cracknell 1973 a).

6.7. *Excitons*

In turning to the study of excitons we pass from the realm of single particle, or quasiparticle, states to more complicated forms of excitations involving more than one particle. An exciton is an excited electronic state of a crystal; it consists of a bound state of an electron–hole pair, resulting from the Coulomb electrostatic attraction between the electron and the hole. In metallic solids any change in the electronic charge distribution resulting from the excitation of one electron will quickly be screened out by a rapid re-arrangement of the other conduction electrons. Thus excitons will be expected to occur only in insulating or semiconducting solids and not in metals. It is easiest to visualize an exciton in a simple ionic insulator such as NaCl or one of the other alkali halides. In a simple-minded approach the electronic states can be regarded as localized states which, to a good approximation, are the same as the states of free Na^+ and Cl^- ions. If an electron wanders away from the ion to which it belongs it leaves behind a hole and the electron–hole pair form bound states. These bound states, which are somewhat similar to the states of a hydrogen atom, are excitons and their existence can be detected in spectroscopic measurements. The idea of an exciton arising from localized electron and hole states is credited to Frenkel (1931). Suppose that an electron leaves a Cl^- ion $(1s^2 2s^2 2p^6 3s^2 3p^6)$ and attaches itself to one of the Na^+ ions $(1s^2 2s^2 2p^6)$ and that the electron goes into one of the lowest available states in the Na^+ ion and came from one of the highest occupied states in the Cl^- ion. The electron and hole states involved in the formation of the exciton will then be a Na^+ $3s$ state and a Cl^- $3p$ state. Strictly speaking, even the bound states of electrons in atoms or ions in a solid should be treated as Bloch states, $\exp(i\mathbf{k} \cdot \mathbf{r})u_{\mathbf{k}}(\mathbf{r})$, and it is the periodic function $u_{\mathbf{k}}(\mathbf{r})$ which we are taking to be the wave function of an electron in a free ion. Even though it is constructed from localized states, an exciton will not remain fixed in any given unit cell for ever but it can move through the crystal. If one abandons the picture of localized states and regards the electrons as being in extended states, which will be more appropriate for semiconductors and also for some insulators, the exciton is still a bound state of an electron–hole pair. But now the wave functions of the electron and the hole are wave functions for extended states. The fundamental excitation is now that of an electron from the full valence band to the empty conduction band, leaving behind a hole in the valence band. The exciton is then a bound state of these extended electron and hole states and is a little more difficult to visualize than in the case of localized states. The idea of the construction of exciton states in this manner from band electron and hole states is credited to Wannier (1937).

The application of group theory to the study of excitons in insulating crystals, such as NaCl, has been considered by Overhauser (1956). Although the excitons are regarded as involving localized Cl^- $3p$ hole states and localized Na^+ $3s$ states it must nevertheless be remembered that these states are in a crystalline environment (see also §§ 2.1 and 4.1–4.5). Suppose that the hole is localized on one of the Cl^- ions. The symmetry of the local environment of a Cl^- ion is that of the point group $m3m$ (O_h) and so we see from table 2 that, being a p state, the hole state must belong to Γ_4^- of $m3m$ (O_h). We are, for the moment, neglecting spin. The excited electron may be on any of the six nearest-neighbour Na^+ ions which are situated at $(\pm \frac{1}{2}a, 0, 0)$ $(0, \pm \frac{1}{2}a, 0)$ and $(0, 0, \pm \frac{1}{2}a)$ relative to the Cl^- ion. The wave function of the electron will therefore be some linear combination of the $3s$ functions which are localized on these six Na^+ ions. We denote these six functions by x, x', y, y', z and z', where x and x' are on opposite sides of the Cl^- ion, etc. These six states form the basis of a reducible representation of $m3m$ (O_h) with characters which can easily be determined by applying the operations of $m3m$ (O_h) :

E	$C_{3j}{}^{\pm}$	C_{2m}	$C_{4m}{}^{\pm}$	C_{2p}
6	0	2	2	0

By inspection of table 2 we see that, on reduction, this gives $\Gamma_1^+ \oplus \Gamma_3^+ \oplus \Gamma_4^-$. The linear combinations of the six Na^+ $3s$ states belonging to these irreducible representations can be constructed quite easily ; they are

$$\Gamma_1^+ : \quad s = (1/\sqrt{6})(x + x' + y + y' + z + z')$$

$$\Gamma_3^+ : \quad u = (1/\sqrt{3})(z + z' - \tfrac{1}{2}(x + x' + y + y'))$$

$$v = \tfrac{1}{2}(x + x' - y - y')$$

$$\Gamma_4^- : \quad p = (1/\sqrt{2})(x - x')$$

$$q = (1/\sqrt{2})(y - y')$$

$$r = (1/\sqrt{2})(z - z')$$

$$(6.65)$$

The number and degeneracy of the energy levels for the excitons can be determined from the reduction of the Kronecker product of the representations to which the Cl^- $3p$ state and the combinations of Na^+ $3s$ states belong ; this reduction can be performed with the aid of table 2 when we obtain

$$\Gamma_4^- \boxtimes (\Gamma_1^+ \oplus \Gamma_3^+ \oplus \Gamma_4^-) =$$
$$\Gamma_1^+ \oplus \Gamma_3^+ \oplus \Gamma_4^+ \oplus \Gamma_5^+ \oplus 2\Gamma_4^- \oplus \Gamma_5^-$$
$$(1) \quad (2) \quad (3) \quad (3) \quad 2(3) \quad (3)$$

$$(6.66)$$

The numbers in brackets below the right-hand side of eqn. (6.66) indicate the degeneracies of the various levels ; including all the degeneracies there is a total of 18 exciton levels altogether.

If we consider the inclusion of spin, but neglect spin–orbit coupling for the moment, the spin of the electron and the spin of the hole will combine to form either a singlet state (with $S = 0$) or a triplet state (with $S = 1$). From table 2 we see that, for our example, the singlet state belongs to Γ_1^+ of $m3m$ and the triplet state belongs to Γ_4^+. If the space parts of the exciton wave functions are combined with the singlet state spin function the resulting wave functions will belong to $\Gamma_4^- \boxtimes (\Gamma_1^+ \oplus \Gamma_3^+ \oplus \Gamma_4^-) \boxtimes \Gamma_1^+$; the reduction of this product will be identical with the right-hand side of eqn. (6.66) namely

$$
\left.
\begin{aligned}
&\Gamma_4^- \boxtimes (\Gamma_1^+ \oplus \Gamma_3^+ \oplus \Gamma_4^-) \boxtimes \Gamma_1^+ = \\
&\quad \Gamma_1^+ \oplus \Gamma_3^+ \oplus \Gamma_4^+ \oplus \Gamma_5^+ \oplus 2\Gamma_4^- \oplus \Gamma_5^- \\
&\quad \;\;(1)\;\;\;\;(2)\;\;\;\;(3)\;\;\;\;(3)\;\;\;\;2(3)\;\;\;\;(3)
\end{aligned}
\right\}. \tag{6.67}
$$

If the space parts of the exciton wave functions are combined with the triplet state spin functions the resulting wave function will belong to $\Gamma_4^- \boxtimes (\Gamma_1^+ \oplus \Gamma_3^+ \oplus \Gamma_4^-) \boxtimes \Gamma_4^+$; this product can be reduced with the aid of table 2 giving

$$
\left.
\begin{aligned}
&\Gamma_4^- \boxtimes (\Gamma_1^+ \oplus \Gamma_3^+ \oplus \Gamma_4^-) \boxtimes \Gamma_4^+ = \\
&\quad \Gamma_1^+ \oplus \Gamma_2^+ \oplus 2\Gamma_3^+ \oplus 4\Gamma_4^+ \oplus 3\Gamma_5^+ \oplus 2\Gamma_1^- \oplus \Gamma_2^- \oplus 3\Gamma_3^- \oplus 3\Gamma_4^- \oplus 3\Gamma_5^- \\
&\quad \;\;(1)\;\;\;\;(1)\;\;\;\;2(2)\;\;\;\;4(3)\;\;\;\;3(3)\;\;\;\;2(1)\;\;\;\;(1)\;\;\;\;3(2)\;\;\;\;3(3)\;\;\;\;3(3)
\end{aligned}
\right\}. \tag{6.68}
$$

The inclusion of spin–orbit coupling would mix singlet and triplet exciton states having the same symmetry. The right-hand sides of eqns. (6.67) and (6.68) indicate the degeneracies that are expected to be caused among the 72 exciton levels as a result of the local symmetry in the crystal. The most common way to observe the existence and effects of excitons is by (optical) spectroscopy. The symmetry assignments of the excitons are then very useful in determining selection rules for the participation of excitons in such spectroscopic experiments (see § 7.2).

We have only considered the details of the determination of the symmetries of excitons when the exciton states are regarded as being constructed out of localized states. Thus we have only determined the symmetries for the excitons at the central point, Γ, of the Brillouin zone. These symmetries are identified by the representations of a point group which are, of course, related in a trivial manner to the $\mathbf{k} = \mathbf{0}$ representations of the space group of the crystal. The arguments can be extended to using localized electron and hole states for the construction of exciton states with non-zero wave vectors. The symmetry assignments and the degeneracies of the exciton states for lines of symmetry passing through Γ, can be obtained from the results for Γ and the compatibility tables for the space group in question. For the other special wave vectors in the Brillouin zone the exciton symmetries can be determined quite conveniently by studying the transformation properties of the operator

$$
a_{\mathbf{k}}^{\dagger} = N^{-1/2} \sum_j \exp (i\mathbf{k} \cdot \mathbf{r}_j) a_j^{\dagger} \tag{6.69}
$$

where $a_j^{\dagger}$ is the operator which creates the excited electron state and the corresponding hole state at site j in the crystal. The arguments can also be extended to the construction of exciton states from non-localized, or band, electron and hole states; this will involve the use of Kronecker products of space-group representations (see § 7.1).

6.8. *Magnetoelastic waves and magnon–phonon interactions*

So far we have seen that, ever since the classic paper by Bouckaert *et al.* (1936), group theory has been used extensively in connection with the electronic energy bands of crystalline solids, while more recently the ideas involved have begun to be applied to other particles, or quasiparticle excitations, in crystalline solids, such as phonons and magnons. Recently the effect of magnon–phonon interactions on the dispersion relations for these quasiparticle excitations in magnetic crystals has come to be of interest as a result of refinements in the techniques involved in inelastic neutron scattering experiments (see, for example, Jensen (1971)). From such experiments it became apparent that, at least for the special lines of symmetry in the Brillouin zone, a given magnon may interact appreciably with certain phonons but not with others. It is possible to investigate these interactions group theoretically (see, for example, Cracknell (1973 c, 1974 c), Montgomery and Cracknell (1973)).

Traditionally it is common to assume that, to a very good first approximation, the lattice vibrations of a crystal are unaffected by the behaviour of any magnetic moments which may be associated with the atoms or ions in the crystal. Similarly, in attempting to describe the motions of the spins in a magnetically ordered system, that is in discussing spin waves, it is also traditional to assume that the atoms or ions with which the magnetic moments are associated are frozen in their equilibrium positions and not involved in any translational motions. This approximation is in the same spirit as the *adiabatic* or *Born–Oppenheimer* approximation which is commonly used when calculating the electronic band structure of a crystalline solid ; in this approximation it is assumed that the electronic band structure can be calculated neglecting the motions of the nuclei, or ions, and assuming them to be frozen in their equilibrium positions. In the non-interacting approximation for magnons and phonons it is possible to represent both the magnon dispersion relations and the phonon dispersion relations on a common diagram, as shown schematically in fig. 18 in which, for simplicity, we just show one acoustic phonon branch AD and one magnon branch BC.

If the non-interacting approximation is abandoned, the situation illustrated by the curves AC and BD in fig. 18 may arise. The Hamiltonian of the system will consist of a lattice part and a spin part and at least one term which involves both the displacements of the nuclei and the orientations of the spins. The collective excitations of the system described by this Hamiltonian are therefore neither pure lattice vibrations nor pure spin waves, but rather they are *magnetoelastic waves*. A magnetoelastic wave involves displacements both in the positions of the atoms in a crystal and in the orientations of the spin vectors associated with the atoms. The existence of such magnetoelastic waves has been known, at least from a theoretical point of view, for a long time (see, for example, chapter 4 of the book by Akhiezer *et al.* (1968)). However, it is only recently that it has become possible to study the dispersion relations for such magnetoelastic waves experimentally. Instead of the intersecting curves AD and BC describing lattice vibrations and spin waves respectively in fig. 18, the two curves no longer cross at $\mathbf{k}_0$ and two non-intersecting curves AC and BD are obtained, each of which

Fig. 18

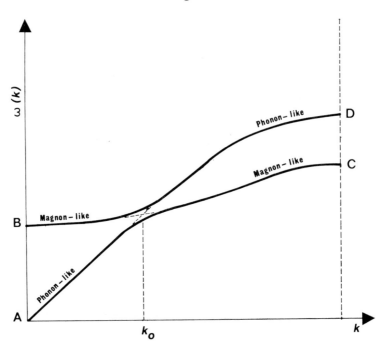

Sketch of the dispersion relations for coupled magnetoelastic waves in a hypothetical magnetic crystal involving one acoustic phonon branch and one
magnon branch. The broken lines indicate the dispersion relations (AD for
phonons and BC for magnons) in the absence of any coupling between spin
waves and lattice vibrations.

describes a magnetoelastic wave. For very small **k**, the quasiparticle excitation described by AC is almost entirely a pure lattice vibration while for very
large **k** near the Brillouin-zone boundary it is almost a pure spin wave; it
is only in the vicinity of $\mathbf{k}_0$ that this quasiparticle excitation involves substantial displacements both in the atomic positions and in the spin orientations.

The same general theorem, namely Wigner's theorem, which lies behind
the use of group theory in labelling the electronic band structure, the phonon
dispersion relations, and the magnon dispersion relations, also applies to
magnetoelastic waves. That is to say, if $\mathscr{H}$ is the Hamiltonian used to
describe the coupled magnetoelastic vibrations of a magnetically-ordered
crystal, any given magnetoelastic wave must belong to one or other of the
irreducible representations, or co-representations, of the group of $\mathscr{H}$, which
in turn will be determined by the symmetry of the crystal itself.

The question of the existence of accidental degeneracies for the dispersion
relations of a particle, or of quasiparticle excitations, in a crystalline solid
was studied by Herring (1937), who showed that the occurrence of an accidental
degeneracy between two branches at the same **k** and belonging to the same
representation is vanishingly improbable (except for certain very special

simple forms of Hamiltonian, such as for example that used in the 'empty-lattice' approximation). The importance of this result has long been appreciated in connection with electronic band structures. Along a line of symmetry we therefore expect to find intersections of electronic energy bands, that is accidental degeneracies between energy bands at certain $\mathbf{k}$, for energy bands belonging to different irreducible representations, but we do not expect to find intersections between two bands belonging to the same representation. If we consider the electronic energy bands along an arbitrary (i.e. non-symmetry) direction of $\mathbf{k}$ from the centre to the boundary of the Brillouin zone, there is only one irreducible space-group representation, $\Gamma_1{}^{\mathbf{k}}$, and all the bands therefore belong to $\Gamma_1{}^{\mathbf{k}}$. This means that for an arbitrary direction of $\mathbf{k}$ we should expect to find no intersections among the energy bands. These arguments were originally given for electronic energy bands and, in the past, they have principally been used in this connection. However, the arguments are general and can be applied equally well to the dispersion relations for any particle or quasiparticle excitation in a crystalline solid. In particular we are concerned at present with their relevance to magnetoelastic waves. Thus, for an arbitrary (i.e. non-symmetry) direction of $\mathbf{k}$ there will only be one irreducible space-group representation at $\mathbf{k}$, $\Gamma_1{}^{\mathbf{k}}$, and all the magnetoelastic waves will belong to this representation ; for an arbitrary direction of $\mathbf{k}$ we therefore expect to find a set of non-intersecting branches for the dispersion relations of the magnetoelastic waves. For a special line of symmetry, however, there will normally be several different irreducible space-group representations at $\mathbf{k}$, $\Gamma_1{}^{\mathbf{k}}$, $\Gamma_2{}^{\mathbf{k}}$, $\Gamma_3{}^{\mathbf{k}}$, etc. and we therefore expect to find a set of dispersion relations for the magnetoelastic waves which have intersections, that is accidental degeneracies at certain $\mathbf{k}$, between branches of different symmetries.

It is common to regard magnetoelastic waves as arising from a relatively weak coupling between the lattice vibrations and the spin waves ; this suggests that it would be profitable to consider the symmetries of coupled magnetoelastic waves in terms of the symmetries of those phonons and magnons which become coupled together. We first note that since the dispersion curve for a magnetoelastic wave, or for any other quasiparticle excitation, is a quasi-continuous function of $\mathbf{k}$, this dispersion curve must belong to the same irreducible representation for all values of $\mathbf{k}$ along a straight line between Γ, the centre of the Brillouin zone, and a point on the boundary of the Brillouin zone. The situation illustrated in fig. 18 applies for an arbitrary (i.e. non-symmetry) direction of $\mathbf{k}$ in the Brillouin zone. All the phonon branches in this particular direction belong to $\Gamma_1{}^{\mathbf{k}}$ and all the magnon branches also belong to $\Gamma_1{}^{\mathbf{k}}$. Similarly all the magnetoelastic waves will also belong to $\Gamma_1{}^{\mathbf{k}}$. Therefore, in this direction any given magnon branch which intersects some phonon branch can interact with it to form a coupled magnetoelastic wave. This means that there is no selection rule preventing the interaction of phonons and magnons for such an arbitrary direction in the Brillouin zone.

We now consider the case of a special line of symmetry in the Brillouin zone. Each branch of the dispersion relations for coupled magnetoelastic waves is a quasi-continuous function of $\mathbf{k}$. Thus, everywhere along this line of symmetry, any given branch of these dispersion relations will be labelled by the same representation, or co-representation, of the space group of the crystal. This leads to the selection rule :

If **k** *is on a line of symmetry the degeneracy between a magnon and a phonon can only be lifted if the magnon and the phonon have the same symmetry, that is they must belong to the same irreducible representation or co-representation, of the space group of the crystal.*

For the purpose of this selection rule we do not regard the two members of a pair of complex conjugate representations as 'different'. In practice the statement of this selection rule may need modification by replacing the word 'different' by the word 'incompatible'; this is because for any given magnetic crystal it is quite common to use different space groups in describing the phonon symmetries and the magnon symmetries. For the phonon symmetries it is common to use the (grey) space group which describes the symmetry of the atomic positions in the crystal and neglects the magnetic moments, whereas for the magnon symmetries it is common to use the magnetic space group or, perhaps, the spin-space group of the magnetically ordered crystal.

Let us suppose that group theory has been used to determine which irreducible representations describe the phonons and the magnons in a given crystal and also to determine the forms of their eigenvectors (see §§ 6.3 and 6.4). Once this has been done it will be possible to see by inspection whether any given magnon branch which intersects a certain phonon branch in the non-interacting approximation will be expected to couple, or mix, with that phonon branch when magnon–phonon interactions are included. Alternatively, if the phonon or magnon symmetries have not been distributed unambiguously among the observed branches of the dispersion relations, the study of the interactions among the various phonon and magnon branches may be useful to assist in determining these assignments unambiguously.

The above discussion of magnon–phonon interactions and the symmetries of magnetoelastic waves has been applied to the examples of ferromagnetic h.c.p. metals (Cracknell 1974 c) and of antiferromagnetic FeF_2 (Cracknell 1973 c). We shall briefly summarize the results for FeF_2. This material has the same crystal structure and configuration of magnetic moments as occurs in antiferromagnetic MnF_2. The magnon symmetries for this structure have already been identified (see fig. 1 and table 27). If the ordering of the magnetic moments is ignored the structure of this material would be the rutile structure which belongs to the classical space group $P4_2/mnm$. The Brillouin zone for this structure was illustrated in fig. 6. The phonon symmetries have been identified for the rutile structure in table 26. In particular the acoustic phonon modes belong to Γ_2^- and Γ_5^- of $P4_2/mnm$ where the notation is that of Dimmock and Wheeler (1962 a). With the aid of the compatibility tables given by Dimmock and Wheeler it is possible to re-assign the phonons to the irreducible representations of $Pnnm$, the unitary sub-group of $P4_2'/mnm'$. Recalling that the symmetries of the magnons have already been identified in table 27 one finds that the interactions between the magnons and the acoustic phonons in FeF_2 should obey the rules given in table 29, where the bracket is used to indicate that Λ_3 and Λ_4 become degenerate when the anti-unitary operations in $P4_2{'}/mnm'$ are included, that is, they form a case (c) co-representation $\mathrm{D}\Lambda_{3,\,4}$ of $P4_2{'}/mnm'$.

The selection rules that we have described so far are independent of the mechanism that is postulated as being responsible for the interaction. Interactions between lattice vibrations and spin waves may arise from several

Table 29.   Selection rules for magnon–(acoustic)phonon interactions along Δ, Σ and Λ in antiferromagnetic FeF_2

$\Delta([100])$: Magnon			Δ_3	Δ_4
Phonon				
LA	Δ_1	Δ_1	$\times$	$\times$
TA	Δ_2	Δ_2	$\times$	$\times$
TA	Δ_3	Δ_3	$\checkmark$	$\times$
$\Sigma([110])$: Magnon			Σ_2	
Phonon				
LA	Σ_1	Σ_1	$\times$	
TA	Σ_2	Σ_1	$\times$	
TA	Σ_4	Σ_2	$\checkmark$	
$\Lambda([001])$: Magnon			Λ_3	Λ_4
Phonon				
LA	Λ_1	Λ_1	$\times$	
TA	Λ_5	$\left.\begin{array}{l}\Lambda_3 \\ \\ \Lambda_4\end{array}\right\}$	$\checkmark$	

Notes.   (i). The magnons are labelled according to the irreducible representations of *Pnnm*.   The phonons are labelled as LA or TA and also according to $P4_2/mnm$ (in column 2) and according to *Pnnm* (in column 3).   (ii). The bracket } or $\smile$ is used to indicate that Λ_3 and Λ_4 become degenerate when the anti-unitary operations of $P4_2{'}/mnm'$ are included.   (Cracknell 1973 c.)

different physical origins ; these include (i) magnetostriction, (ii) changes in the crystalline electric field, (iii) a more complicated interaction proposed by Liu (1972) involving the annihilation of a phonon and the creation of a magnon via the (virtual) excitation of an electron–hole pair.   One can find in the literature various different forms for the terms to include in the Hamiltonian if one wishes to describe magnon–phonon interactions.   The simplest form of interaction can be written as

$$\mathscr{H}_{mp}{'} = \sum_{\mathbf{k}} U(\mathbf{k})(\alpha_{\mathbf{k}}\beta_{\mathbf{k}}{}^{\dagger} + \alpha_{\mathbf{k}}{}^{\dagger}\beta_{\mathbf{k}}) \tag{6.70}$$

(see p. 74 of Kittel (1963)).   In eqn. (6.70) $\alpha_{\mathbf{k}}^{\dagger}$ and $\beta_{\mathbf{k}}^{\dagger}$ are creation operators for magnons and phonons respectively, and $U(\mathbf{k})$ is a function representing the magnitude of the interaction.   A typical term in eqn. (6.70) such as

$$V_1 = U(\mathbf{k})\alpha_{\mathbf{k}}^{\dagger}\beta_{\mathbf{k}} \tag{6.71}$$

corresponds to the annihilation of a phonon with wave vector $\mathbf{k}$ and the creation of a magnon with wave vector $\mathbf{k}$. When considering magnon–phonon interactions in Tb, Jensen (1971) obtained a more complicated expression for the interaction in which a typical term was of the form

$$V_2 = U(\mathbf{k}_2, \mathbf{k}_1)\alpha_{\mathbf{k}_1+\mathbf{k}_2}^{\dagger}\alpha_{\mathbf{k}_1}\beta_{\mathbf{k}_2} \tag{6.72}$$

which corresponds to the creation of one magnon with wave vector $\mathbf{k}_1 + \mathbf{k}_2$ by annihilating one magnon with wave vector $\mathbf{k}_1$ and one phonon with wave vector $\mathbf{k}_2$, or, in other words, the scattering of a magnon from $\mathbf{k}_1$ to $(\mathbf{k}_1 + \mathbf{k}_2)$ with the absorption of a phonon with wave vector $\mathbf{k}_2$. $U(\mathbf{k}_2, \mathbf{k}_1)$ represents the magnitude of the interaction.   Various other forms for the interaction terms have also been considered (see, for example, Huberman and Burstein (1971)).   Using the theory of the Kronecker products of space group representations (see § 7.1) it is possible to show that an interaction of the form in eqn. (6.71) leads to the same selection rule as that which was given earlier in this section as a result of general arguments.   It is also possible to show that the more complicated interactions, such as the one given in eqn. (6.72) may lead to additional selection rules ; for further details see § 5 of the paper by Cracknell (1974 c).

6.9. *Other ' -ons '*

We have described those applications of group theory which have already actually been used extensively for quasiparticles in solids.   But we have by no means exhausted all the quasiparticles which are known to solid-state physicists and to which group theory could, in principle, be applied.   The list could be extended to include polaritons, plasmons, polarons and magnetic polarons (or spin polarons or magnarons) and possibly some others too. Although very little work has so far been published in this connection, the use of group theory could also be extended to some of these other quasi-particles too.   In this section we shall consider what lines this extension might take.   Descriptions of the physical natures of several of these quasi-particles will be found in the book by Sherwood (1972).

A polaron consists of a moving electron, usually in an ionic crystal, which is surrounded by a cloud of virtual phonons or, in more pictorial terms, a moving electron accompanied by an associated local crystal deformation. Other types of polaron are also possible in which the electron is surrounded by a cloud of some other quasiparticles instead of phonons.   Thus we have the possibility of an excitonic polaron, if the accompanying cloud of quasi-particles surrounding the electron consists of excitons, and we have the possibility of a magnetic polaron in a magnetically ordered material, if the surrounding cloud of quasiparticles consists of magnons.   It would not be appropriate to include a lengthy discussion of polaron theory here (for details see, for example, Kuper and Whitfield (1963)).   We simply note that the formation of a polaron involves phonons, or other appropriate quasiparticles, with all possible wave vectors in the Brillouin zone. Since the general wave vectors vastly outnumber the wave vectors of special symmetry, and since group theory can only usefully be applied to the wave vectors of special symmetry, it would seem that group theory is of no particular relevance in connection with polaron formation.

Group theory could be applied to the discussion of polaritons in a very similar manner to the discussion of magnetoelastic waves that we gave in § 6.8. A polariton is a quantum of energy of a coupled wave which is

Fig. 19

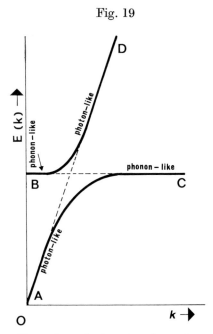

Sketch of the polariton dispersion relations in a hypothetical crystal, involving only one phonon branch. The broken lines indicate the dispersion relations (AD for photons and BC for phonons) in the absence of any phonon–photon coupling.

analogous to a magnetoelastic wave, but with the difference that for the polariton the quasiparticles involved in the coupling are a photon and a phonon instead of a magnon and a phonon. Thus for polariton formation we can draw a picture similar to fig. 18 with photon and (optical) phonon branches which would cross in the absence of any interaction but which may separate when the coupling is included (see fig. 19) to form two non-intersecting coupled-mode, or polariton, branches when the interaction is included. The situation may actually be more complicated than fig. 19 would suggest, because of the presence of both longitudinal and transverse phonons and also, at least in crystals of finite size, because of the excitation of surface modes. It would not be appropriate here to give a lengthy discussion of the theory and properties of polaritons (see, for example, Ruppin and Englman (1970), Barker and Loudon (1972), Sherwood (1972)) ; we shall restrict our attention to their group-theoretical aspects.

For a wave vector along an arbitrary direction passing through Γ, $\mathbf{k}$ has no special symmetry and there will be no selection rule preventing the phonon–photon coupling from leading to coupled polariton modes, assuming that the phonon modes are neither pure longitudinal nor pure transverse modes (see § 6.3). However, if the wave vector $\mathbf{k}$ corresponds to a special line of symmetry there will be several possible irreducible representations $\Gamma_1{}^{\mathbf{k}}$, $\Gamma_2{}^{\mathbf{k}}$, ... $\Gamma_j{}^{\mathbf{k}}$, ... of the space group $\mathbf{G}$ of the crystal. Therefore, by similar arguments to those used previously for magnetoelastic waves, along a line of symmetry we only expect the separation shown in fig. 19 to occur if the phonon and photon both belong to the same irreducible representation, $\Gamma_j{}^{\mathbf{k}}$, of $\mathbf{G}$. If they belong to different representations then we would expect the phonon and photon branches to cross, without any separation. Then, just as in § 6.8, it is possible to use group theory to see whether any given phonon branch which intersects a certain photon branch, in the non-interacting approximation, will be expected to couple, or mix, with that photon branch when phonon–photon interactions are included. There remains, of course, the problem of identifying the phonon symmetries and the photon symmetries. The problem of the identification of the phonon symmetries has been discussed in § 6.3. The identification of the irreducible representation to which a photon is assigned will depend on the form of the coupling that one assumes to be responsible for causing the polariton formation. Thus if the phonon–photon coupling were supposed to involve the electric field vectors of the photons, the irreducible representations used for the photons will be the ones to which a polar vector belong.

In addition to considering group-theoretical restrictions on the possible phonon–photon pairs which are allowed to couple to form polaritons, it is also possible to consider whether polariton branches will couple with further phonon branches. Moreover, one can also investigate the possibility of the intersection of two polariton branches. The latter possibility has been treated algebraically by Lamprecht and Merten (1969) but not (as far as I know) group-theoretically. Along a line of symmetry one may have a crossing of polariton branches with different group-theoretical labels (see fig. 20).

Fig. 20

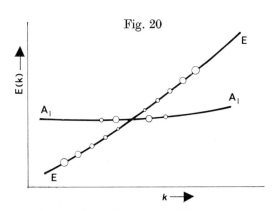

Sketch of crossing polariton branches with different symmetries, adapted from Claus (1970). *Note added in proof.*—The paper by Claus (1970) should be read in conjunction with a more recent paper (W. Nitsch and R. Claus, ' Interaction of isotropic modes with polaritons ', *Z. Naturf.* A (in the press).

But for an arbitrary direction of **k** this crossing will be removed, because all the polaritons must belong to the same irreducible representation, and gaps will appear instead of the crossing. This behaviour has been observed in $LiNbO_3$ (Winter and Claus 1972, Posledovich *et al.* 1973).

The arguments we have outlined for polaritons could also, presumably, be adapted for plasmaritons which arise from the coupling of phonons with plasmons (instead of with photons); a brief discussion of plasmaritons will be found in the book by Sherwood (1972).

§ 7. SELECTION RULES

In this section we shall be concerned with selection rules for a collection of several phenomena involving the interactions of particles or quasiparticles in a solid with other particles or quasiparticles that originate from, or radiate to, the region of space surrounding the solid. Thus we shall be concerned with the absorption or scattering of beams of particles or beams of photons when they fall on a specimen of a crystalline solid. The particles involved are either neutrons or electrons. We have already mentioned the use of the elastic scattering of X-rays, electrons, and neutrons in structure determination in § 1.2 ; the elastic scattering enables the mean positions of the atoms, or the mean orientations of the magnetic moments of the atoms, in a crystal to be determined. In addition to the elastic scattering there is also the possibility that incident particles or photons may be scattered inelastically, that is with the transfer of energy to or from the crystal.

The inelastic scattering enables one to probe the behaviour of itinerant (conduction) electrons or of quasiparticle excitations in a crystal. Because of the very high photon energy of an X-ray photon, relative to a typical phonon or magnon energy, the use of inelastic X-ray scattering (sometimes called ' thermal diffuse X-ray scattering ') is rather inconvenient. The use of inelastic neutron scattering has been particularly successful for point by point mapping of the phonon dispersion relations and magnon dispersion relations (see §§ 6.3, 6.4 and 7.3). It should be possible, in principle, to determine the phonon or magnon dispersion relations by using inelastic electron scattering. In the case of low-energy electrons the range of the electrons is so small (see table 1) that the particle or quasiparticle states in the solid that are involved in the interactions with the electrons will be surface states rather than bulk states of the solid. Therefore, the use of the inelastic scattering of electrons should enable detailed studies of surface vibrations and surface spin waves to be undertaken ; indeed, successful results have recently been obtained along these lines for a few materials. It is partly because the techniques in the preparation and characterization of the surface of a solid are rather difficult that low-energy electron scattering has only recently come to be studied to any great extent. From the theoretical point of view, the study of surface electronic energy states and of surface phonons and magnons has also lagged behind the study of the corresponding states or excitations for bulk materials. Inelastic neutron scattering is likely to be difficult to use for the study of surface excitations because of the very low scattering cross sections for neutrons (Rieder and Hörl 1968). Another important feature results from the large difference between the magnitudes of the absorption cross sections for inelastic neutron scattering and inelastic electron scattering. Whereas each neutron in a scattered neutron beam is very unlikely to have been scattered more than once, this is not so for low-energy electron scattering for which multiple scattering will occur.

As well as the study of phonons and magnons there is also the possibility of the use of inelastic scattering of electrons in studying electronic states in crystals. At first sight it might appear, by analogy with inelastic neutron scattering, that the inelastic scattering of electrons would provide a convenient way to obtain a point by point determination of the band structure, that is the curves of $E(\mathbf{k})$ against $\mathbf{k}$ for the valence and conduction electrons, of a solid. However, for a number of reasons this is not the case (see § 7.4).

By the term 'selection rules' in this section we mean extra restrictions which apply to interactions for particles or quasiparticles with certain wave vectors, but which would not apply for particles or quasiparticles with other wave vectors. We have already encountered a related problem in § 6.8, namely the determination of selection rules for the coupling of two quasiparticles (a magnon and a phonon) with a common wave vector and a common energy in a crystalline solid.

7.1. Kronecker products and selection rules

The theoretical study of an absorption or scattering cross section, for a process involving the absorption or scattering of particles or quasiparticles with wave vectors $\mathbf{k}_i$ which impinge on a crystal, involves the calculation of transition probabilities which in turn involves the calculation of quantum-mechanical matrix elements. It is possible to regard the absorption or scattering in terms of processes involving the creation or annihilation of particles or quasiparticles with wave vectors $\mathbf{k}_j$ in the crystal. Then the transition probability can be expressed in terms of matrix elements that involve particle, or quasiparticle, wave functions and appropriate interaction parameters or coupling constants $V(\mathbf{k}_i, \mathbf{k}_j)$. Such a matrix element would take the form

$$\langle \psi_q{}^{\mathbf{k}_j} V \psi_p{}^{\mathbf{k}_i} \rangle = \int \psi_q{}^{\mathbf{k}_j}(\mathbf{r})^* V(\mathbf{k}_i, \mathbf{k}_j) \psi_p{}^{\mathbf{k}_i}(\mathbf{r}) \, d\mathbf{r} \qquad (7.1)$$

where $\psi_p{}^{\mathbf{k}_i}(\mathbf{r})$ and $\psi_q{}^{\mathbf{k}_j}(\mathbf{r})$ are the wave functions of the particles or quasiparticles involved in the process. For the simplest possible process, the absorption of a quasiparticle with wave vector $\mathbf{k}_i$ and the creation of a single quasiparticle with wave vector $\mathbf{k}_j$, the wave functions $\psi_p{}^{\mathbf{k}_i}(\mathbf{r})$ and $\psi_q{}^{\mathbf{k}_j}(\mathbf{r})$ will be single quasiparticle wave functions. However, the incident quasiparticle may be absorbed by a more complicated process, or the incident particle or quasiparticle may be scattered instead of being absorbed. In these situations the wave functions $\psi_p{}^{\mathbf{k}_i}(\mathbf{r})$ and $\psi_q{}^{\mathbf{k}_j}(\mathbf{r})$ may be compound wave functions for two, or even more, particles or quasiparticles. It is also possible that the scattering process may involve a sequence of transitions, instead of a single transition ; in this case the transition probability will not involve just a single matrix element $\langle \psi_q{}^{\mathbf{k}_j} V \psi_p{}^{\mathbf{k}_i} \rangle$ but will involve a product of two or more matrix elements. Because of the many different particle or quasiparticle wave vectors $\mathbf{k}_j$ available in a solid, the absorption cross section or scattering cross section for a given incident particle or quasiparticle state will involve a summation of matrix elements, or products of matrix elements, over all possible wave vectors that contribute to the scattering.

The core of the group-theoretical treatment of these selection rules is then concerned with studying quantum-mechanical matrix elements of the form in eqn. (7.1) for a system with the symmetry of a certain space group, $\mathbf{G}$, or magnetic space group, $\mathbf{M}$. Then one will be concerned with the use of both Neumann's principle and Wigner's theorem (see § 1.3). Because the matrix element is related to a physical observable, namely a transition probability, the matrix element must possess all the symmetry of the Hamiltonian, $\mathscr{H}_0$, of the crystal. In other words, the matrix element must belong to the identity representation Γ_1 (or Γ_1^+) of the group $\mathbf{G}$ of the Hamiltonian $\mathscr{H}_0$. For the moment we consider only unitary, or classical, space groups, $\mathbf{G}$. However, the wave functions $\psi_p^{\mathbf{k}_i}(\mathbf{r})$ and $\psi_q^{\mathbf{k}_j}(\mathbf{r})$ will not necessarily possess all the symmetry of $\mathbf{G}$ and $\mathscr{H}_0$, but will belong to irreducible representations $\Gamma_p^{\mathbf{k}_i}$ and $\Gamma_q^{\mathbf{k}_j}$, respectively, of $\mathbf{G}$ which are not, in general, Γ_1 (or Γ_1^+). Moreover, the symmetry of the interaction potential, or coupling constant, $V(\mathbf{k}_i, \mathbf{k}_j)$ will not necessarily be the same as that of $\mathbf{G}$ and $\mathscr{H}_0$. There are two possibilities to be considered; the first is when the symmetry of the interaction potential $V(\mathbf{k}_i, \mathbf{k}_j)$ is the same or higher than the symmetry of $\mathscr{H}_0$. In this case $V(\mathbf{k}_i, \mathbf{k}_j)$ belongs to Γ_1 (or Γ_1^+) of $\mathbf{G}$, so that the matrix element $\langle \psi_q^{\mathbf{k}_j} V \psi_p^{\mathbf{k}_i} \rangle$ belongs to the Kronecker product $\Gamma_q^{\mathbf{k}_j *} \boxtimes \Gamma_p^{\mathbf{k}_i}$. In order to satisfy Neumann's principle this product must contain the identity representation Γ_1 (or Γ_1^+) of $\mathbf{G}$ or else the interaction will be forbidden. Therefore the interaction involving the two states $\psi_p^{\mathbf{k}_i}(\mathbf{r})$ and $\psi_q^{\mathbf{k}_j}(\mathbf{r})$ will be forbidden by symmetry unless they belong to the same irreducible representation of $\mathbf{G}$. The second possibility is that the interaction potential $V(\mathbf{k}_i, \mathbf{k}_j)$ has a lower symmetry than the symmetry of $\mathbf{G}$. In this case the interaction may consist of various terms that belong to different irreducible representations of $\mathbf{G}$. For simplicity we assume that $V(\mathbf{k}_i, \mathbf{k}_j)$ belongs to a particular irreducible representation $\Gamma_r^{\mathbf{k}_k}$ of $\mathbf{G}$. The matrix element $\langle \psi_q^{\mathbf{k}_j} V \psi_p^{\mathbf{k}_i} \rangle$ will then belong to the triple Kronecker product $\Gamma_q^{\mathbf{k}_j *} \boxtimes \Gamma_r^{\mathbf{k}_k} \boxtimes \Gamma_p^{\mathbf{k}_i}$. Therefore, invoking Neumann's principle, unless this product contains the identity representation Γ_1 (or Γ_1^+) of $\mathbf{G}$ the matrix element must vanish and the interaction will be forbidden. This imposes restrictions on the representations $\Gamma_p^{\mathbf{k}_i}$ and $\Gamma_q^{\mathbf{k}_j}$ which can lead to a non-vanishing matrix element and this, in turn, imposes selection rules in a number of physical situations, as we shall see later in this section.

The basic mathematical manipulation which has to be achieved is the reduction of the Kronecker product of two irreducible representations of a space group. We recall that the wave function of a particle or quasiparticle with wave vector $\mathbf{k}$ belongs to an induced representation $(\Gamma_i^{\mathbf{k}} \uparrow \mathbf{G})$ of $\mathbf{G}$, where $\Gamma_i^{\mathbf{k}}$ is a representation of the little group $\mathbf{G}^{\mathbf{k}}$ which is a subgroup of $\mathbf{G}$. The problem then is to determine the coefficients $C_{pq,\,r}^{\mathbf{k}_i \mathbf{k}_j,\,\mathbf{k}_l}$ in the Clebsch–Gordan decomposition

$$(\Gamma_p^{\mathbf{k}_i} \uparrow \mathbf{G}) \boxtimes (\Gamma_q^{\mathbf{k}_j} \uparrow \mathbf{G}) \equiv \sum_r \sum_l C_{pq,\,r}^{\mathbf{k}_i \mathbf{k}_j,\,\mathbf{k}_l} (\Gamma_r^{\mathbf{k}_l} \uparrow \mathbf{G}). \qquad (7.2)$$

The theory involved in the determination of the coefficients $C_{pq,\,r}^{\mathbf{k}_i \mathbf{k}_j,\,\mathbf{k}_l}$ is quite complicated and it would not be appropriate to describe it here; for details see, for example, § 4.7 of Bradley and Cracknell (1972) and the references cited therein. We simply note that there are two stages in the determination of $C_{pq,\,r}{}^{\mathbf{k}_i\mathbf{k}_j,\,\mathbf{k}_l}$. The first involves the wave vectors $\mathbf{k}_i$, $\mathbf{k}_j$ and $\mathbf{k}_l$. If we suppose that $\mathbf{k}_i$ and $\mathbf{k}_j$ are given then $\mathbf{k}_l$ is not free to take all values in the Brillouin zone but the possible values of $\mathbf{k}_l$ are very restricted. $C_{pq,\,r}{}^{\mathbf{k}_i\mathbf{k}_j,\,\mathbf{k}_l}$ vanishes for all values of $\mathbf{k}_l$ except those for which

$$R_\beta\mathbf{k}_i + R_\alpha\mathbf{k}_j \equiv \mathbf{k}_l \tag{7.3}$$

where R_α and R_β are the rotational parts of some, but not necessarily all, of the operations $\{R|\mathbf{v}\}$ of the space group $\mathbf{G}$. The details involved in identifying R_α and R_β do not concern us here (see, for example, p. 211 of Bradley and Cracknell (1972)). In practice the restrictions on α and β are so severe that there is rarely more than one set of α and β so that eqn. (7.3) frequently leads to only one value of $\mathbf{k}_l$ for which the coefficients $C_{pq,\,r}{}^{\mathbf{k}_i\mathbf{k}_j,\,\mathbf{k}_l}$ do not automatically vanish. A restriction imposed on $\mathbf{k}_l$, for given $\mathbf{k}_i$, $\mathbf{k}_j$ and $\mathbf{G}$, is sometimes described as a *wave vector selection rule* (WVSR). Having identified the rather small number of wave vectors $\mathbf{k}_l$ for which $C_{pq,\,r}{}^{\mathbf{k}_i\mathbf{k}_j,\,\mathbf{k}_l}$ does not automatically vanish it remains to use a standard formula to evaluate $C_{pq,\,r}{}^{\mathbf{k}_i\mathbf{k}_j,\,\mathbf{k}_l}$ (see, for example, eqn. (4.7.29) of Bradley and Cracknell (1972)) for each value of r where $\mathbf{k}_i$, $\mathbf{k}_j$, $\mathbf{k}_l$, p and q are given.

To emphasize the point that more than one value of $\mathbf{k}_l$ does sometimes occur we quote the results of an example given by Bradley (1966) using the

Table 30. Character tables for Γ and M of $P23$

$\mathbf{G}^\Gamma$	E	$3C_{2m}$	$4C_{3j}{}^+$	$4C_{3j}{}^-$
A	1	1	1	1
^1E	1	1	ω^*	ω
^2E	1	1	ω	ω^*
T	3	-1	0	0

$\mathbf{G}^M$	E	C_{2x}	C_{2y}	C_{2z}
A_1	1	1	1	1
B_1	1	-1	-1	1
B_2	1	-1	1	-1
B_3	1	1	-1	-1

Notes. (i) $\omega = \exp(2\pi i/3)$. (ii) All translations in $\mathbf{G}^\Gamma$ are represented by the identity. (iii) In $\mathbf{G}^M$, $\{E|\mathbf{t}_1\}$ and $\{E|\mathbf{t}_2\}$ are represented by -1 and $\{E|\mathbf{t}_3\}$ is represented by $+1$.

Table 31. The inner Kronecker products of representations belonging to M for $P23$

$$MA_1 \boxtimes MA_1 = \Gamma A \oplus \Gamma^1 E \oplus \Gamma^2 E \oplus 2MA_1$$
$$MA_1 \boxtimes MB_1 = \Gamma T \oplus MB_2 \oplus MB_3$$
$$MA_1 \boxtimes MB_2 = \Gamma T \oplus MB_3 \oplus MB_1$$
$$MA_1 \boxtimes MB_3 = \Gamma T \oplus MB_1 \oplus MB_2$$
$$MB_1 \boxtimes MB_1 = \Gamma A \oplus \Gamma^1 E \oplus \Gamma^2 E \oplus 2MB_1$$
$$MB_1 \boxtimes MB_2 = \Gamma T \oplus MB_3 \oplus MA_1$$
$$MB_1 \boxtimes MB_3 = \Gamma T \oplus MB_2 \oplus MA_1$$
$$MB_2 \boxtimes MB_2 = \Gamma A \oplus \Gamma^1 E \oplus \Gamma^2 E \oplus 2MB_2$$
$$MB_2 \boxtimes MB_3 = \Gamma T \oplus MB_1 \oplus MA_1$$
$$MB_3 \boxtimes MB_3 = \Gamma A \oplus \Gamma^1 E \oplus \Gamma^2 E \oplus 2MB_3$$

cubic space group $P23$ (T^1). The Brillouin zone for this space group is illustrated in fig. 5 (*a*). We consider the reduction of products of pairs of representations of $P23$ induced from representations with wave vectors corresponding to M, the mid-point of an edge of the Brillouin zone. That is $\mathbf{k}_i = M$ and $\mathbf{k}_j = M$. The irreducible representations of the little group $\mathbf{G}^M$ are identified in table 30. It is possible to show in this example that there are two wave vectors $\mathbf{k}_l$ which do not automatically lead to the vanishing of $C_{pq,\,r}{}^{\mathbf{k}_i \mathbf{k}_j,\,\mathbf{k}_l}$, namely Γ and M. The reductions of the Kronecker products of all the pairs of (single-valued) representations of $P23$ for $\mathbf{k}_i = \mathbf{k}_j = M$ are listed in table 31, where the representations of $\mathbf{G}^\Gamma$ and $\mathbf{G}^M$ are identified in table 30.

In this section we have been careful to refer to the reduction of products of representations of a space group $\mathbf{G}$. The representations of $\mathbf{G}$ are the induced representations ($\Gamma_p{}^{\mathbf{k}_i} \uparrow \mathbf{G}$) where $\Gamma_p{}^{\mathbf{k}_i}$ are the (small) representations of the little group $\mathbf{G}^{\mathbf{k}_i}$. $\mathbf{G}^{\mathbf{k}_i}$ is a subgroup, albeit possibly a trivial subgroup, of $\mathbf{G}$. The approach we have indicated is sometimes called the *full-group method* and has principally been used by Birman (1962 b). The justification for using it is that, because the Hamiltonian of the system possesses the full symmetry of the space group $\mathbf{G}$, we know from the use of Wigner's theorem that the wave function of a particle or quasiparticle must belong to some irreducible representation of $\mathbf{G}$. These representations of $\mathbf{G}$ are, of course, determined and labelled by using the (small) representations of the little groups $\mathbf{G}^{\mathbf{k}}$ (see § 5.1). In practice, however, because of the undoubted importance of $\mathbf{G}^{\mathbf{k}}$ it is quite common to speak of the wave function of a particle or quasiparticle in a solid as belonging to one of the (small) representations of the little group $\mathbf{G}^{\mathbf{k}}$. From this point of view it then seems natural to try to study the symmetry properties of a product of two wave functions $\psi_p{}^{\mathbf{k}_i}$ and $\psi_q{}^{\mathbf{k}_j}$ by considering the reduction of the Kronecker product of $\Gamma_p{}^{\mathbf{k}_i}$ and $\Gamma_q{}^{\mathbf{k}_j}$. This approach is called the *subgroup method* and it has been used by a number of authors (Elliott and Loudon 1960, Lax and Hopfield 1961, Lax 1962, 1965, Folland and Bassani 1968, Doni and Pastori Parravicini 1973). The fact that $\mathbf{G}^{\mathbf{k}_i}$ and $\mathbf{G}^{\mathbf{k}_j}$ may be different groups leads to a formal difficulty in that in trying to form the product $\Gamma_p{}^{\mathbf{k}_i} \boxtimes \Gamma_q{}^{\mathbf{k}_j}$ one is trying to form a Kronecker product of two different groups. Initially doubts were expressed about the completeness of the subgroup method and several workers attempted to reconcile the full-group method and the subgroup method (Lax 1965, Birman 1966 a, Zak 1966); the complete justification of the subgroup method in terms of the full-group method, as well as an exposition of some of the rather subtle differences between the two methods, was given in a paper by Bradley (1966). In practice, whether one uses the full-group method or the subgroup method should be determined by considering which is the more convenient for the particular material and problem under consideration. A practical comparison between the full-group and subgroup methods has been given for the diamond structure, that is space group $Fd3m$ ($O_h{}^7$), by Davies and Lewis (1971).

A special case of the Kronecker product of two space-group representations of the form given in eqn. (7.2) arises when $p = q$ and $\mathbf{k}_i = \mathbf{k}_j$; in this case we have the square of the representation ($\Gamma_p{}^{\mathbf{k}_i} \uparrow \mathbf{G}$). When discussing the Kronecker products of point-group representations in § 1.5 we noted that, in addition to squares, cubes, or nth powers, of point-group representations, it was also possible to define and identify symmetrized and antisymmetrized powers of point-group representations. In a similar manner it is possible to define and identify symmetrized and antisymmetrized powers of space-group representations, although the manipulation is naturally much more complicated for space groups than for point groups (for details of the general theory see Bradley and Davies (1970), Lewis (1973), Gard (1973 a, b, c)). As in the evaluation of a Kronecker product of the form in eqn. (7.2), there are two stages; the first stage involves the determination of the wave vector selection rules (WVSRs) to identify the wave vectors which can appear in the reduction, while in the second stage one determines how many times each of the representations, for each of these allowed wave vectors, is actually present in the reduction. The only structure for which complete results have been determined for the symmetrized and antisymmetrized squares of the space group representations appears to be the zinc-blende structure which belongs to the space group $F\bar{4}3m$ (T_d^2) (Birman 1962 a, b, 1963, Karavaev 1966, Bradley and Davies 1970). So far the cubes and higher powers have received little attention as far as the application of the general theory to particular space groups is concerned.

7.2. *Infra-red absorption, Raman scattering, and magnon sidebands*

If infra-red radiation impinges on a crystal it is possible for a photon of the radiation to be absorbed with the excitation of one or more phonons in the crystal provided both the law of conservation of energy and the law of conservation of momentum are satisfied. The momentum of the photon can be expressed as $2\pi\hbar/\lambda$, where λ is the wavelength. Compared with the momentum of a phonon in the crystal this is very small, since a typical value of the phonon momentum is of the order of $2\pi\hbar/a$, where a is the interatomic spacing. On the scale of the Brillouin zone of a crystal the momentum of the infra-red photon is therefore practically zero, so that to conserve momentum the absorption of an infra-red photon can either create a single phonon at $\mathbf{k} = \mathbf{0}$, that is, at the central point Γ of the Brillouin zone, or else two phonons with equal and opposite momenta, that is, one at $\mathbf{k}$ and one at $-\mathbf{k}$. The conservation of energy is satisfied for the one-phonon process by having the energy of the infra-red photon equal to the energy of the phonon of the lattice vibration which is excited at Γ, that is

$$\hbar\omega_0 = \hbar\omega_1(\mathbf{0}). \qquad (7.4)$$

where $\omega_0/2\pi$ is the frequency of the infra-red photon. The occurrence of this one-phonon absorption process is restricted to ionic crystals. For the two-phonon process the energy of the infra-red photon is equal to the total energy of the two phonons produced, so that

$$\hbar\omega_0 = \hbar\omega_1(\mathbf{k}) + \hbar\omega_2(-\mathbf{k}). \qquad (7.5)$$

Assuming the phonon spectrum at $-\mathbf{k}$ is the same as that at $\mathbf{k}$, eqn. (7.5) simplifies to

$$\hbar\omega_0 = \hbar\omega_1(\mathbf{k}) + \hbar\omega_2(\mathbf{k}). \qquad (7.6)$$

If one uses a source of infra-red radiation containing a continuous spread of frequencies one would expect to find absorption of the infra-red radiation for the whole range of phonon frequencies in the crystal. Prominent features in this absorption curve may occur when $\omega_1(\mathbf{k})$ and $\omega_2(\mathbf{k})$ in eqn. (7.6) correspond to phonons with wave vectors $\mathbf{k}$ of special points of symmetry in the Brillouin zone. This is because of variations in the selection rules governing the two-phonon absorption process for different wave vectors $\mathbf{k}$. If the radiation excites two phonons on the same branch of the dispersion relations then $\omega_1(\mathbf{k})$ and $\omega_2(\mathbf{k})$ in eqn. (7.6) will be equal. Infra-red absorption measurements do not, of course, measure $\mathbf{k}$ and the value of $\mathbf{k}$ has to be inferred. It is also possible for an infra-red photon to be absorbed with the creation of more than two phonons, but the probability of such a process occurring is very small. For a crystal which is magnetically ordered an incident infra-red photon can be absorbed with the creation of either one or two magnons. The whole of the discussion given earlier in this paragraph could then be repeated but with the substitution of the word ' magnon ' for the word ' phonon '. The details of selection rules for infra-red absorption for wave vectors corresponding to special points of symmetry will, however, be different for the case of magnons from the case of phonons.

Instead of the absorption of a photon by the creation of phonons or of magnons, there is also the possibility of inelastic scattering of photons (Raman scattering or Brillouin scattering) with the creation of phonons or magnons. The photon frequencies involved correspond to optical frequencies and there has been a considerable upsurge of interest in Raman scattering by solids following the invention of the laser. As one might expect, the special points of symmetry may give rise to sharp features in the spectrum of the scattered radiation. Selection rules can be deduced for Raman scattering involving phonons or magnons at special points of symmetry in the Brillouin zone. If we are concerned with magnetically ordered crystals the spin waves can also manifest their existence by giving rise to magnon sidebands on sharp optical absorption lines. Yet again one can expect the magnons at the special points of symmetry to be particularly important and selection rules can be deduced for this phenomenon.

We consider first the determination of selection rules for infra-red absorption involving the creation of only one phonon ; this involves a $\mathbf{k} = \mathbf{0}$ phonon which we shall suppose belongs to the representation Γ_λ. Since the $\mathbf{k} = \mathbf{0}$ representations of a space group are isomorphic with the irreducible representations of some point group, $\mathbf{G}$, the determination of these selection rules is quite straightforward. We consider the operator W corresponding to an electric dipole moment or a magnetic dipole moment. W then transforms like a polar vector (x, y, z) for electric dipole absorption or like an axial vector (S_x, S_y, S_z) for magnetic dipole absorption. In practice the probability of magnetic dipole absorption is very small. It is possible to assign the components of the operator W to the appropriate irreducible representations of each of the 32 classical point groups (see, for example, Eyring *et al.* (1944) or Koster *et al.* (1963)). The three irreducible representations $\Gamma_p{}^{\mathbf{k}_i}$, $\Gamma_q{}^{\mathbf{k}_j}$ and $\Gamma_r{}^{\mathbf{k}_k}$ to which the various factors in the matrix element $\langle \psi_q{}^{\mathbf{k}_j} V \psi_p{}^{\mathbf{k}_i} \rangle$ belong, therefore describe the following :

$\Gamma_p{}^{\mathbf{k}_i}$ initial state of solid, no phonon excited, Γ_1 (or $\Gamma_1{}^+$),

$\Gamma_q{}^{\mathbf{k}_j}$ final state of solid, one phonon excited, Γ_λ,

$\Gamma_r{}^{\mathbf{k}_k}$ electric or magnetic dipole operator, Γ_W.

Therefore this one-phonon process will be allowed if $\Gamma_\lambda \boxtimes \Gamma_W$ contains the identity representation Γ_1 (or Γ_1^+) of the point group, **G**. The reduction of $\Gamma_\lambda \boxtimes \Gamma_W$ can be obtained with the aid of the tables of Koster *et al.* (1963). For example, the $\mathbf{k}=\mathbf{0}$ phonons in the diamond structure belong to Γ_{15} and Γ_{25}', see fig. 17, where the notation is that of Herring (1942). In this structure the operator W transforms in the following manner (see, for example, table 87 of Koster *et al.* (1963)):

$$\text{electric dipole } (x, y, z) \qquad \Gamma_{15},$$
$$\text{magnetic dipole } (S_x, S_y, S_z) \qquad \Gamma_{15}'.$$

The reductions of the products $\Gamma_\lambda \boxtimes \Gamma_W$ which have to be considered can be obtained, for example, from the tables given by Birman (1962 b):

$$\left.\begin{array}{l}
\Gamma_{15} \boxtimes \Gamma_{15} = \Gamma_1 \oplus \Gamma_{12} \oplus \Gamma_{15}' \oplus \Gamma_{25}' \\[2ex]
\Gamma_{15} \boxtimes \Gamma_{15}' = \Gamma_1' \oplus \Gamma_{12}' \oplus \Gamma_{15} \oplus \Gamma_{25} \\[2ex]
\Gamma_{25}' \boxtimes \Gamma_{15} = \Gamma_2' \oplus \Gamma_{12}' \oplus \Gamma_{15} \oplus \Gamma_{25} \\[2ex]
\Gamma_{25}' \boxtimes \Gamma_{15}' = \Gamma_2 \oplus \Gamma_{12} \oplus \Gamma_{15}' \oplus \Gamma_{25}'
\end{array}\right\}. \qquad (7.7)$$

From these reductions we see that the Γ_{15} phonons (the acoustic phonons at $\mathbf{k}=\mathbf{0}$) are infra-red active for electric dipole absorption but not for magnetic dipole absorption and the Γ_{25}' phonons are inactive for both electric dipole absorption and magnetic dipole absorption. In addition to indicating whether a particular $\mathbf{k}=\mathbf{0}$ phonon is able to take part in one-phonon absorption of infra-red radiation, it also indicates the allowed polarization for the photon which is absorbed; this is obtained from the assignments of the individual components of (x, y, z) or (S_x, S_y, S_z) to the irreducible representations of **G**.

It is also possible to determine selection rules for infra-red absorption with the creation of two phonons. The conservation of momentum requires that the two phonons which are created have equal and opposite wave vectors $\mathbf{k}_m$ and $-\mathbf{k}_m$. Suppose that these phonons belong to the irreducible representations $\Gamma_\mu^{\mathbf{k}_m}$ and $\Gamma_\nu^{-\mathbf{k}_m}$, where we have to allow the possibility that μ and ν may not be the same. The difference from the one-phonon case is that the final state wave function $\psi_q^{\mathbf{k}_j}$ is a two-phonon wave function which belongs to $\Gamma_\mu^{\mathbf{k}_m} \boxtimes \Gamma_\nu^{-\mathbf{k}_m}$. It is, therefore, necessary to obtain the reduction of this Kronecker product to determine the symmetry of the two-phonon wave function. Once the representation to which the two-phonon wave function belongs has been identified, it can be used in place of Γ_λ in the one-phonon treatment which we have already described. The details of the determination of the selection rules are therefore more complicated than for the one-phonon case. Consider the phonons which have wave vectors corresponding to the point X in the Brillouin zone for the diamond structure. They belong to the representations X_1, X_3 and X_4, see fig. 17, where the labels are in the notation of Herring (1942). The possible two-phonon states, with one phonon at $\mathbf{k}$ and the other at $-\mathbf{k}$, therefore belong to Γ_λ which may be the product of any two of these representations at X. The product $\Gamma_\lambda \boxtimes \Gamma_W$ is therefore a triple Kronecker product $X_i \boxtimes X_j \boxtimes \Gamma_{15}$ for electric-dipole absorption involving a pair of phonons belonging to X_i and X_j; for this pair of phonons to be able to participate in electric dipole absorption this triple product must contain Γ_1. The reductions of these triple products can be determined by repeated application of the tables given by Birman (1962 b). The complete analysis would be somewhat lengthy and so we just give one example. From those tables we see that

$$X_1 \boxtimes X_3 = \Gamma_{15} \oplus \Gamma_{15}' \oplus \Gamma_{25} \oplus \Gamma_{25}' \oplus X_1 \oplus X_2 \oplus X_3 \oplus X_4. \tag{7.8}$$

Therefore since $X_1 \boxtimes X_3$ contains Γ_{15}, the triple product $X_1 \boxtimes X_3 \boxtimes \Gamma_{15}$ will contain Γ_1 and so the creation of a pair of phonons belonging to X_1 and X_3, by infra-red absorption, is allowed as an electric dipole process. The procedure could be repeated for magnetic dipole processes by using Γ_{15}' as W. There is a complication we have not mentioned so far. If the two phonons, one at $\mathbf{k}$ and the other at $-\mathbf{k}$, belong to the same representation say X_i, then it is the symmetrized square $[X_i]^2$, rather than $X_i \boxtimes X_i$ which should be used as Γ_λ. Selection rules for the special points of symmetry in a number of important space groups have been derived by several authors (Birman 1962 b, 1963, Gorzkowski 1964, Burstein *et al.* 1965, Birman *et al.* 1966, Cornwell 1966, Olbrychski and Van Huong 1970).

Similar arguments to those which we have just described for the absorption of infra-red photons with the excitation of phonons can be adapted to the absorption of optical photons with the formation of excitons. We have already discussed the identification of the symmetries of the excitons in § 6.7. The above arguments can also be repeated for infra-red absorption involving the creation of one magnon or of two magnons. The only difference from the previous argument would be that all the wave functions and the operator W should be assigned to the irreducible co-representations of a magnetic space group $\mathbf{M}$ instead of to the irreducible representations of a classical space group, $\mathbf{G}$. For a one-magnon process it is a $\mathbf{k} = \mathbf{0}$ magnon which is involved, so that this would simplify to the consideration of the irreducible co-representations of one of the black and white, or magnetic, point groups. For a two-magnon process it will involve the Kronecker product of two co-representations $D\Gamma_\mu{}^{\mathbf{k}_m} \boxtimes D\Gamma_\nu{}^{-\mathbf{k}_m}$. However, if one is only concerned with a 'yes' or 'no' type answer to the question of whether a particular process is forbidden or not, this can be obtained by using only the unitary subgroup $\mathbf{H}$ of $\mathbf{M}$ instead of $\mathbf{M}$ itself.

We illustrate the theory of infra-red absorption with the creation of either one or two magnons by considering the example of antiferromagnetic MnF_2 (Loudon 1968). The symmetries of the magnons at the various points of symmetry in the Brillouin zone of antiferromagnetic MnF_2 have already been identified (see table 27). The components of the electric dipole operator transform like x, y and z which belong to Γ_3^-, Γ_4^- and Γ_2^-, respectively, of *Pnnm*, the unitary subgroup, or to $D\Gamma_{3,4}^-$ and $D\Gamma_2^-$ of $P4_2'/mnm'$. The components of the magnetic dipole operator transform like S_x, S_y and S_z which belong to $D\Gamma_{3,4}^+$ and $D\Gamma_2^+$, respectively. The active states and the polarizations of the photons involved are summarized in table 32. Since the magnons at $\mathbf{k} = \mathbf{0}$ in antiferromagnetic MnF_2 belong to $D\Gamma_{3,4}^+$, infra-red absorption with the creation of one magnon cannot occur as an electric dipole process and can only occur as a magnetic dipole process with infra-red photons polarized with the magnetic vector normal to the c axis. We have already

Table 32. Active states at $\mathbf{k}=0$ for electric and magnetic dipole absorption in MnF$_2$, FeF$_2$ or CoF$_2$

	Active state
Electric dipole E_x, E_y (σ and α) E_z (π)	$D\Gamma_{3,\,4}^{-}$ $D\Gamma_2^{-}$
Magnetic dipole L_x, L_y (π and α) L_z (σ)	$D\Gamma_{3,\,4}^{+}$ $D\Gamma_2^{+}$

noted that for the two-magnon process the two magnons which are created are at $\mathbf{k}$ and $-\mathbf{k}$. For any given wave vector $\mathbf{k}$ there are two types of magnon corresponding to $\alpha_{\uparrow\mathbf{k}}^{\dagger}$ and $\alpha_{\downarrow\mathbf{k}}^{\dagger}$. Therefore the two-magnon absorption process involves either $\alpha_{\uparrow\mathbf{k}}^{\dagger}$ or $\alpha_{\downarrow\mathbf{k}}^{\dagger}$ and either $\alpha_{\uparrow-\mathbf{k}}^{\dagger}$ or $\alpha_{\downarrow-\mathbf{k}}^{\dagger}$. The four different zero wave-vector two-magnon states can be denoted by

$$\left.\begin{aligned}
|0,\,-\rangle &= |\uparrow\mathbf{k},\,\downarrow-\mathbf{k}\rangle - |\downarrow\mathbf{k},\,\uparrow-\mathbf{k}\rangle \\
|0,\,+\rangle &= |\uparrow\mathbf{k},\,\downarrow-\mathbf{k}\rangle + |\downarrow\mathbf{k},\,\uparrow-\mathbf{k}\rangle \\
|2,\,+\rangle &= |\uparrow\mathbf{k},\,\uparrow-\mathbf{k}\rangle \\
|-2,\,+\rangle &= |\downarrow\mathbf{k},\,\downarrow-\mathbf{k}\rangle
\end{aligned}\right\}. \tag{7.9}$$

The number 0, $+2$ or -2 indicates the S_z quantum number of each state and the $\pm$ indicates the parity of each state. From eqn. (7.9) we see that, for any wave vector $\mathbf{k}$ in the Brillouin zone we have a selection rule ; only $|0,\,-\rangle$, the state with negative parity, is active in electric-dipole absorption, while only the other three states would be active in magnetic dipole absorption. There may be additional selection rules that apply at the special points of symmetry. One can study the transformation properties of the four two-magnon states in eqn. (7.9) under the operations of $P4_2'/mnm'$ and

Table 33. The symmetries of the two-magnon states for wave vectors corresponding to points of symmetry on the surface of the Brillouin zone of antiferromagnetic MnF$_2$

Point of symmetry	Symmetries of two-magnon states		
X	Γ_1^{+},	Γ_2^{+},	Γ_3^{-}
Z	Γ_1^{+},	Γ_2^{+},	Γ_1^{-}
R	Γ_1^{+},	Γ_2^{+},	Γ_3^{+}
M	Γ_1^{+},	Γ_2^{+},	Γ_2^{+}
A	Γ_1^{+},	Γ_2^{+},	Γ_2^{-}
Y	Γ_1^{+},	Γ_2^{+},	Γ_4^{-}
U	Γ_1^{+},	Γ_2^{+},	Γ_4^{+}

thereby assign these two-magnon states to the appropriate co-representations at Γ. Alternatively, in a similar manner to that used before for phonons, one can form Kronecker products of the co-representations to which the magnons at the points of symmetry belong. Where the magnons belong to a degenerate co-representation it is the symmetrized square of that co-representation which is used. The symmetries of the two-magnon states for the points of symmetry on the surface of the Brillouin zone of antiferro-magnetic MnF_2 are identified in table 33.

The selection rules for the inelastic scattering of photons (Raman scatter-ing or Brillouin scattering) by phonons or by magnons can be obtained in a similar manner to that just described for infra-red absorption. The only important difference is that the operator W is now the electric polarizability α_{ij}. The components of α_{ij} transform like

$$\alpha_{ij} : \begin{pmatrix} x^2 & xy & xz \\ yx & y^2 & yz \\ zx & zy & z^2 \end{pmatrix}.$$

The component α_{ij} that is actually involved for any given phonon or magnon indicates the possible polarizations of the incident and scattered photons. We have to construct a table to replace table 32 for Raman scattering. This is done by assigning the components of α_{ij} to the various $\mathbf{k} = \mathbf{0}$ irreducible co-representations of $P4_2'/mnm'$; these co-representations are isomorphic with the co-representations of the point group $4'/mmm'$. The results are given in table 34 where we have introduced a shorthand notation for the components of α_{ij} and we have not required these components to be real ; the polarizations of the electric fields $\mathbf{E}^1$ and $\mathbf{E}^2$ of the incident and scattered photons are also indicated in table 34. It is only the three even-parity states $|2, +\rangle$, $|-2, +\rangle$ and $|0, +\rangle$ which may be involved in Raman scattering. Suppose, for example, we consider the $|0, +\rangle$ state which is thought to be the important one. Since the $\mathbf{k} = \mathbf{0}$ magnons belong to $D\Gamma_{3,4}^+$ we see from table 34 that the allowed combinations of electric field of the incident and scattered photons are $E_x^1 E_z^2$, $E_y^1 E_z^2$, $E_z^1 E_x^2$ and $E_z^1 E_y^2$. Thus for incident light that is linearly polarized with its electric field vector parallel to the c axis (E_z^1) the one-magnon scattering produces light that has its electric field vector in the ab plane (E_x^2, E_y^2). Using table 34 and the symmetries of the two-magnon states given in table 33, selection rules for two-magnon Raman scattering can be written down in a similar manner.

The theory of the determination of selection rules for magnon sidebands on optical absorption or emission spectra resembles the theory for infra-red absorption involving phonons or magnons. The practical differences arise from the fact that the exciton energies are very much larger than the magnon energies. If a photon is completely absorbed with the formation of an exciton, the conservation of momentum requires that this is a $\mathbf{k} = \mathbf{0}$ exciton and the selection rules can be found in the same way as for infra-red absorption with the creation of one phonon. If a photon is absorbed with the creation of one exciton and one magnon, the exciton and the magnon must have equal and opposite wave vectors, $\mathbf{k}$ and $-\mathbf{k}$, to conserve momentum. One can

Table 34.   Active states for Raman scattering in MnF_2, FeF_2 or CoF_2

α_{ij}	Photon polarizations	Active state
$\begin{pmatrix} a & 0 & 0 \\ 0 & a^* & 0 \\ 0 & 0 & c \end{pmatrix}$	$E_x{}^1E_x{}^2,\ E_y{}^1E_y{}^2,\ E_z{}^1E_z{}^2$	$D\Gamma_1{}^+$
$\begin{pmatrix} 0 & d & 0 \\ d^* & 0 & 0 \\ 0 & 0 & 0 \end{pmatrix}$	$E_x{}^1E_y{}^2,\ E_y{}^1E_x{}^2$	$D\Gamma_2{}^+$
$\begin{pmatrix} 0 & 0 & f \\ 0 & 0 & 0 \\ g & 0 & 0 \end{pmatrix}$	$E_x{}^1E_z{}^2,\ E_z{}^1E_x{}^2$	$D\Gamma_{3,\,4}{}^+$
$\begin{pmatrix} 0 & 0 & 0 \\ 0 & 0 & -f^* \\ 0 & -g^* & 0 \end{pmatrix}$	$E_y{}^1E_z{}^2,\ E_z{}^1E_y{}^2$	

(Fleury and Loudon (1968), Cracknell 1969 b))

construct two-quasiparticle states, analogous to those in eqn. (7.9), but where each state now involves one exciton and one magnon. Once again we consider the example of a simple antiferromagnet with the rutile structure.

Exciton lines have been observed in antiferromagnetic FeF_2 with all the positive-parity symmetries $D\Gamma_1{}^+$, $D\Gamma_2{}^+$ and $D\Gamma_{3,\,4}{}^+$, where the labels refer to co-representations of the magnetic space group $P4_2{}'/mnm'$; these labels are derived from those used by Dimmock and Wheeler (1962 a) for $Pnnm$, the unitary subgroup of $P4_2{}'/mnm'$. Suppose that we consider electric-dipole absorption involving the exciton bands that belong to $D\Gamma_1{}^+$ and $D\Gamma_2{}^+$ at $\mathbf{k}=\mathbf{0}$; the group theoretical labels for these exciton bands at the other points of symmetry and along the lines of symmetry are given in table 35. We denote these exciton states, for arbitrary $\mathbf{k}$, by $|1\mathbf{k}\rangle$ and $|2\mathbf{k}\rangle$, respectively. We have to excite one of these two excitons at $\mathbf{k}$, or $-\mathbf{k}$, and one of the four magnons $\alpha_{\uparrow\mathbf{k}}{}^\dagger$, $\alpha_{\downarrow\mathbf{k}}{}^\dagger$, $\alpha_{\uparrow-\mathbf{k}}{}^\dagger$ and $\alpha_{\downarrow-\mathbf{k}}{}^\dagger$, subject to the condition that the exciton and the magnon have opposite wave vectors $\mathbf{k}$ and $-\mathbf{k}$ or vice versa. There are, therefore, eight exciton–magnon states which can be constructed with zero wave vector. Four of these have positive parity and cannot contribute to electric-dipole absorption while the remaining four exciton–magnon states which have negative parity can be written as

$$\left.\begin{aligned}
|a\rangle &= |1\mathbf{k},\,\uparrow-\mathbf{k}\rangle - |1-\mathbf{k},\,\uparrow\mathbf{k}\rangle + |1\mathbf{k},\,\downarrow-\mathbf{k}\rangle - |1-\mathbf{k},\,\downarrow\mathbf{k}\rangle \\
|b\rangle &= |1\mathbf{k},\,\uparrow-\mathbf{k}\rangle - |1-\mathbf{k},\,\uparrow\mathbf{k}\rangle - |1\mathbf{k},\,\downarrow-\mathbf{k}\rangle + |1-\mathbf{k},\,\downarrow\mathbf{k}\rangle \\
|c\rangle &= |2\mathbf{k},\,\uparrow-\mathbf{k}\rangle - |2-\mathbf{k},\,\uparrow\mathbf{k}\rangle + |2\mathbf{k},\,\downarrow-\mathbf{k}\rangle - |2-\mathbf{k},\,\downarrow\mathbf{k}\rangle \\
|d\rangle &= |2\mathbf{k},\,\uparrow-\mathbf{k}\rangle - |2-\mathbf{k},\,\uparrow\mathbf{k}\rangle - |2\mathbf{k},\,\downarrow-\mathbf{k}\rangle + |2-\mathbf{k},\,\downarrow\mathbf{k}\rangle
\end{aligned}\right\} \quad (7.10)$$

where the notation is that of Loudon (1968). The symbol $|1\mathbf{k},\,\uparrow-\mathbf{k}\rangle$, for example, indicates the combination of the $\Gamma_1{}^+$ exciton band (at $\mathbf{k}$) with the $\alpha_{\uparrow-\mathbf{k}}{}^\dagger$ magnon (at $-\mathbf{k}$). By studying the transformation properties of the exciton–magnon states in eqn. (7.10) under the symmetry operations of the magnetic space group $P4_2'/mnm'$ it is possible to assign these exciton–magnon states to the appropriate space-group representations at Γ. The results when $\mathbf{k}$, or $-\mathbf{k}$, is one of the points of symmetry in the Brillouin zone are

$$
\begin{array}{ccccc}
 & X & Z & A & Y \\
|a\rangle + |d\rangle & \Gamma_1^- & \Gamma_4^- & \Gamma_3^- & \Gamma_2^- \\
|b\rangle + |c\rangle & \Gamma_2^- & \Gamma_3^- & \Gamma_4^- & \Gamma_1^-
\end{array}
$$

(see Loudon (1968)). The selection rules for electric-dipole absorption can then be obtained with the aid of table 32. For $\mathbf{E}\|c$ only X and Y are electric-dipole active and for $\mathbf{E}\perp c$ only A and Z are electric-dipole active. These selection rules will govern the existence of sharp features that may be exhibited in the magnon sidebands although, of course, the general shape of the sidebands will be governed by the exciton–magnon states arising from excitons and magnons with general wave vectors. Calculations of the complete magnon sideband shape involves postulating a form for the exciton–magnon coupling and knowing the dispersion relations for both the excitons and the magnons ; the details are described in § 7 of the article by Loudon (1968).

Table 35.   Exciton and magnon symmetries in antiferromagnetic MnF_2, FeF_2 or CoF_2

	Excitons	Excitons Magnons
Γ	$D\Gamma_1^+ \oplus D\Gamma_2^+$	$D\Gamma_{3,\,4}^+$
M	$DM_1^+ \oplus DM_2^+$	$DM_3^+ \oplus DM_4^+$
Z	DZ_1	DZ_2
A	DA_1	DA_2
R	DR_1^+	DR_1^+
X	DX_1	DX_2
Δ	$D\Delta_1 \oplus D\Delta_2$	$D\Delta_3 \oplus D\Delta_4$
U	DU_1	DU_1
Λ	$D\Lambda_1 \oplus D\Lambda_2$	$D\Lambda_{3,\,4}$
V	$DV_1 \oplus DV_2$	$DV_3 \oplus DV_4$
Σ	$2D\Sigma_1$	$2D\Sigma_2$
S	$DS_1 \oplus DS_2$	$DS_1 \oplus DS_2$
Y	$DY_1 \oplus DY_2$	$DY_3 \oplus DY_4$
T	DT_1	DT_1
W	DW_1	DW_1

Note. The exciton symmetries in column 2 and the exciton and magnon symmetries in column 3 are labelled in terms of the co-representations of $P4_2'/mnm'$, based on the notation used by Dimmock and Wheeler (1962 a) for the representations of the unitary subgroup *Pnnm*.

7.3. *Neutron scattering*

The inelastic scattering of thermal neutrons provides the principal technique for the experimental investigation of the dispersion relations for both phonons and magnons in crystals. Suppose that an incident neutron with energy E_0 and momentum $\hbar\mathbf{k}_0$ is scattered inelastically, with the creation of one phonon or of one magnon, and that the final energy, E_1, and momentum, $\hbar\mathbf{k}_1$, of the scattered neutron are measured. Then the principle of the

conservation of energy and the principle of the conservation of momentum
lead to equations which enable one to determine the energy and momentum
of the phonon or magnon which was excited, namely

$$E(\mathbf{k}) = E_0 - E_1 \tag{7.11}$$

and

$$\hbar\mathbf{k} = \hbar\mathbf{k}_0 - \hbar\mathbf{k}_1 - \mathbf{G_n}. \tag{7.12}$$

$\mathbf{G_n}$ is a reciprocal lattice vector $n_1\mathbf{g}_1 + n_2\mathbf{g}_2 + n_3\mathbf{g}_3$, where n_1, n_2 and n_3 are
integers. $E(\mathbf{k})$ and $\hbar\mathbf{k}$ identify a point on the appropriate dispersion rela-
tions, and from measurements involving various scattering vectors the
dispersion curves can be obtained. The conditions obtained from the conservation equations are built into the expression for the scattering cross
section. However, in the expression for the scattering cross section there
may be other factors which cause the cross section to vanish even for some
phonons or magnons which are able to satisfy the principles of the conservation of energy and of momentum. The problem of the determination of
selection rules for the inelastic scattering of neutrons by phonons and magnons
has been considered by a number of authors (Elliott and Thorpe 1967, Casella
and Trevino 1972, Cracknell and Sedaghat 1973, Frikkee 1973).

The complete expression for the one-phonon inelastic neutron scattering
cross section is rather complicated (see, for example, § 4.3 of the book by
Marshall and Lovesey (1971)). We are not concerned with the complete
expression at present but only with any factors which depend on the symmetry
of the crystal and which therefore may vanish, leading to the vanishing of
the cross section itself, for phonons with certain symmetries. Suppose that
a phonon with frequency $\omega_j(\mathbf{k})/2\pi$ is created in scattering a neutron from $\mathbf{k}_0$
to $\mathbf{k}_1$. Then the conservation of momentum requires that

$$\mathbf{k}_0 = \mathbf{k}_1 + \mathbf{k} + \mathbf{G_n}. \tag{7.13}$$

It is convenient to introduce the scattering vector

$$\mathbf{K} = \mathbf{k}_1 - \mathbf{k}_0. \tag{7.14}$$

$\mathbf{K}$ is therefore also given by

$$\mathbf{K} = -\mathbf{k} - \mathbf{G_n}. \tag{7.15}$$

The symmetry-dependent factor in the contribution to the one-phonon
scattering cross section arising from a phonon with frequency $\omega_j(\mathbf{k})/2\pi$ is
$|\mathbf{K} . \mathbf{g}_j(\mathbf{K})|^2$. $\mathbf{g}_j(\mathbf{K})$ is the structure factor defined by

$$\mathbf{g}_j(\mathbf{K}) = \sum_\kappa a_\kappa' \mathbf{e}(\kappa|\mathbf{k}j) \exp\{i\mathbf{K} . \mathbf{r}(\kappa)\} \tag{7.16}$$

where $\mathbf{e}(\kappa|\mathbf{k}j)$ is the normal mode eigenvector specifying the displacements
of the atoms labelled by κ in a unit cell in the normal mode specified by j
and $\mathbf{k}$. $\mathbf{e}(\kappa|\mathbf{k}j)$ belongs to one of the irreducible representations, say $\Gamma_r^{\mathbf{k}}$,
of the space group; this representation can be identified by the procedure
outlined in § 6.3.

For a given irreducible representation, $\Gamma_r^{\mathbf{k}}$, the total contribution to the
symmetry-dependent factor in the scattering cross section can be written as

$$F_r(\mathbf{K}) = \sum_j |\mathbf{K} . \mathbf{g}_j(\mathbf{K})|^2 \tag{7.17}$$

where the summation over j is restricted to those phonon branches that belong to this particular irreducible representation. It is then possible to identify the representation, say $\Gamma_s{}^{\mathbf{K}}$, to which the factor $\exp{(i\mathbf{K} \cdot \mathbf{r}(\kappa))}$ in eqn. (7.16) belongs. $\Gamma_s{}^{\mathbf{K}}$ could be identified, for example, by using projection operator techniques. Having identified $\Gamma_r{}^{\mathbf{k}}$ and $\Gamma_s{}^{\mathbf{K}}$ it is possible to form the Kronecker product $\Gamma_r{}^{\mathbf{k}} \boxtimes \Gamma_s{}^{\mathbf{K}}$, which is the representation to which the structure factor $\mathbf{g}_j(\mathbf{K})$ belongs. The symmetry of $F_r(\mathbf{K})$ can then be determined. If the identity representation Γ_1 (or $\Gamma_1{}^+$) at $\mathbf{k} = \mathbf{0}$ is not contained among the representations that appear in the reduction of the representation to which $F_r(\mathbf{K})$ belongs, the scattering of a neutron from $\mathbf{k}_0$ to $\mathbf{k}_1$ by phonons belonging to $\Gamma_r{}^{\mathbf{k}}$ will be forbidden. A slightly different procedure was employed by Elliott and Thorpe (1967). By exploiting the invariance of $F_r(\mathbf{K})$ under the symmetry operations of the space group and using the characters of the elements in the space group, Elliott and Thorpe obtained an expression for $F_r(\mathbf{K})$ in the form

$$F_r(\mathbf{K}) = K^2 \sum_{\beta\gamma} \sum_s (a_s')^2 l^\beta B_s{}^{\beta\gamma}(r, \mathbf{K}) l^\gamma \qquad (7.18)$$

where $\mathbf{l}$ is a unit vector parallel to $\mathbf{K}$, so that $\mathbf{K} = K\mathbf{l}$, and a_s' is defined in eqn. (3) of Elliott and Thorpe (1967). $B_s{}^{\beta\gamma}(r, \mathbf{K})$ is a symmetric tensor for the species of atom labelled by s and this tensor contains all the symmetry-dependent part of the expression for $F_r(\mathbf{K})$. The actual expression for $B_s{}^{\beta\gamma}(r, \mathbf{K})$ is complicated and need not detain us here; for details see eqn. (23) of Elliott and Thorpe (1967). If, for a given representation $\Gamma_r{}^{\mathbf{k}}$ all the components of the tensor $B_s{}^{\beta\gamma}(r, \mathbf{K})$ vanish, this means that phonons belonging to the representation $\Gamma_r{}^{\mathbf{k}}$ are inactive in this neutron scattering. The symmetry properties only determine the total scattering from all the phonons belonging to $\Gamma_r{}^{\mathbf{k}}$.

A detailed analysis to determine which phonons are active and which are inactive in inelastic neutron scattering has been performed by Elliott and Thorpe (1967) for all the special points and lines of symmetry for the fluorite structure (CaF_2, UO_2, etc.) and for cuprite (Cu_2O). An extract from their results for the fluorite structure, which belongs to the space group $Fm3m$ ($O_h{}^5$), is reproduced in table 36, where the phonon symmetries are labelled with the notation used by Bouckaert *et al.* (1936) for this space group. It will be noticed that the tensor $B_s{}^{\beta\gamma}(r, \mathbf{K})$ is tabulated for some values of $\mathbf{K}$ outside the Brillouin zone. This is because, although the structure factor $\mathbf{g}_j(\mathbf{K})$ is periodic, its periodicity need not be the full periodicity of the reciprocal lattice. $\mathbf{g}_j(\mathbf{K})$ needs to be found within a ' structure zone ' which is defined by reciprocal lattice vectors $\mathbf{G}_n'$ such that, for all atoms in the unit cell,

$$\exp{\{i\mathbf{G}_n' \cdot \mathbf{r}(\kappa)\}} = 1. \qquad (7.19)$$

In some crystals, when the $\mathbf{r}(\kappa)$ are not simply related to the lattice vectors the structure zone may be infinite in some directions; this is not the case with the fluorite structure. In the fluorite structure the atoms can be taken to be at the positions

$$\mathbf{r}(1) \quad Ca \quad (0, 0, 0)$$
$$\mathbf{r}(2) \quad F \quad +(\tfrac{1}{4}, \tfrac{1}{4}, \tfrac{1}{4})a$$
$$\mathbf{r}(3) \quad F \quad -(\tfrac{1}{4}, \tfrac{1}{4}, \tfrac{1}{4})a$$

so that, from eqn. (7.19), the structure lattice vectors, $\mathbf{G_n}'$, are $(2\pi/a)(0, 2, 2)$, $(2\pi/a)(2, 0, 2)$, $(2\pi/a)(2, 2, 0)$, etc. In this case it is quite easy to show that the volume of the structure zone is four times the volume of the Brillouin zone. Selection rules can then easily be determined with the aid of table 36. Thus, if all the components of $B_s{}^{\beta\gamma}(r, \mathbf{K})$ are zero, for $\mathbf{K}$ and all $\mathbf{K} + \mathbf{G}'$, the phonons $\Gamma_r{}^{\mathbf{k}}$ are inactive. Thus, for example, the Γ_{25}' phonons are inactive.

Table 36. The components of the tensor $B^{\gamma\delta}(r, \mathbf{K})$ for the fluorite structure

| Point | Phonon | Calcium | | | | | | Fluorine | | | | | |
		B^{11}	B^{22}	B^{33}	B^{12}	B^{13}	B^{23}	B^{11}	B^{22}	B^{33}	B^{12}	B^{13}	B^{23}
$\dfrac{2\pi}{a}(0, 0, 0)$	$2\Gamma_{15},\ \Gamma_{25}'$	1	1	1				2	2	2			
$\dfrac{2\pi}{a}(1, 1, 1)$	$2\Gamma_{15},\ \Gamma_{25}'$	1	1	1				2	2	2			
$\dfrac{2\pi}{a}(q, 0, 0)$	$2\Delta_1,\ \Delta_2'$	1						2					
	$3\Delta_5$		1	1					2	2			
$\dfrac{2\pi}{a}(q+1, 1, 1)$	$2\Delta_1,\ \Delta_2'$	1						2					
	$3\Delta_5$		1	1					2	2			
$\dfrac{2\pi}{a}(1, 0, 0)$	$X_1,\ X_2'$							2					
	X_4'	1											
	X_5								2	2			
	$2X_5'$		1	1									
$\dfrac{2\pi}{a}(2, 1, 1)$	$X_1,\ X_2'$							2					
	X_4'	1											
	X_5												
	$2X_5'$		1	1					2	2			

Note. This is an extract from the paper by Elliott and Thorpe (1967) ; the results for other special wave vectors can be found in that paper. The Brillouin zone is illustrated in fig. 8.

Δ_2' phonons are inactive at $(2\pi/a)(q, 0, 0)$ but are active at $(2\pi/a)(q+1, 1, 1)$. However, even when some of the components of $B_s{}^{\beta\gamma}(r, \mathbf{K})$ are non-zero, $F_r(\mathbf{K})$ may still vanish ; this can occur through the vanishing of $|\mathbf{K} \cdot \mathbf{g}_j(\mathbf{K})|$ resulting from $\mathbf{g}_j(\mathbf{K})$ being perpendicular to $\mathbf{K}$. However, if $\mathbf{K}$ and $\mathbf{g}_j(\mathbf{K})$ are mutually perpendicular there will be some $\mathbf{G_n}'$ such that $\mathbf{K} + \mathbf{G_n}'$ is not perpendicular to $\mathbf{g}_j(\mathbf{K})$. Such branches may be loosely described as ' transverse ', although it is not necessary for all $\mathbf{e}(\kappa|kj)$ to be perpendicular to $\mathbf{K}$. The Δ_5 phonons provide an example of this. They are transverse and inactive at $(2\pi/a)(q, 0, 0)$ in spite of the fact that some components of $B_s{}^{\beta\gamma}(r, \mathbf{K})$ are non-zero ; however the Δ_5 phonons are active at $(2\pi/a)(q, 2, 0)$.

The details of the determination of selection rules for the inelastic scattering of neutrons involving the creation of magnons are simpler than for the corresponding phonon case (Elliott and Thorpe 1967, Cracknell and Sedaghat 1973). As in the case of phonons, the scattering is governed by the principle of the conservation of momentum and the principle of the conservation of energy, see eqns. (7.11) and (7.12). It would not be appropriate here to become involved in a lengthy discussion of the calculation of the cross sections for scattering by magnons (for details see, for example, Halpern and Johnson (1939), Egelstaff (1965), or Marshall and Lovesey (1971)). We also find it convenient to use the scattering vector $\mathbf{K}$ again, see eqns. (7.14) and (7.15). It is possible to show that the cross section for scattering by magnons with wave vector $\mathbf{k}$ contains the square of a structure factor

$$F(\mathbf{K}) = \sum_{\kappa} f_{\kappa}(\mathbf{K}) \exp \{i(\mathbf{K} \pm \mathbf{k}) \cdot \mathbf{r}(\kappa)\}. \tag{7.20}$$

$f_{\kappa}(\mathbf{K})$ is the magnetic form factor for the atom specified by κ and the summations are over all the atoms in one unit cell. $F(\mathbf{K})$ does not depend on the group of $\mathbf{K}$ and therefore does not depend on the representation, or co-representation, of the group of the wave vector of the magnon involved in the scattering. In this respect, therefore, it contrasts with the phonon case, because the selection rules for inelastic neutron scattering by phonons depend on both $\mathbf{k}$ and the irreducible representations (at $\mathbf{k}$) to which the phonons belong. Therefore, in the determination of selection rules for neutron scattering by magnons we have to consider the possibility that for certain $\mathbf{K}$ the structure factor $F(\mathbf{K})$ in eqn. (7.20) vanishes.

For the inelastic scattering of thermal neutrons by a ferromagnetic b.c.c. or f.c.c. metal there is only one atom per unit cell, that is only one value of κ, and we can choose the origin in such a manner that $\mathbf{r}(\kappa) = \mathbf{0}$. Therefore, the exponential factor $\exp \{i(\mathbf{K} + \mathbf{k}) \cdot \mathbf{r}(\kappa)\}$ is equal to unity for all $\mathbf{K}$ so that, for these cases, there is no wave vector $\mathbf{k}$ for which the magnons are forbidden by symmetry to participate in the scattering. For a ferromagnetic h.c.p. metal there are two atoms per unit cell and the study of $F(\mathbf{K})$ is slightly more complicated (Cracknell and Sedaghat 1973). The atoms occupy the Wyckoff position c or d, with coordinates (Henry and Lonsdale 1965)

$$c : \tfrac{1}{3}, \tfrac{2}{3}, \tfrac{1}{4} ; \ \tfrac{2}{3}, \tfrac{1}{3}, \tfrac{3}{4}$$
$$d : \tfrac{1}{3}, \tfrac{2}{3}, \tfrac{3}{4} ; \ \tfrac{2}{3}, \tfrac{1}{3}, \tfrac{1}{4}$$

referred to the hexagonal axes x, y and z with 120° between the $+x$ and $+y$ directions. We choose the sites d and for convenience we move the origin to the point $\tfrac{1}{2}, \tfrac{1}{2}, \tfrac{1}{2}$ when the coordinates of these sites become

$$d : \ -\tfrac{1}{6}, \tfrac{1}{6}, \tfrac{1}{4} ; \ \tfrac{1}{6}, -\tfrac{1}{6}, -\tfrac{1}{4}.$$

The expression for $f(\mathbf{K})$ therefore involves a factor

$$\hat{F}(\mathbf{K}) = 2 \cos \{\mathbf{K} \cdot (\tfrac{1}{6}\mathbf{t}_1 - \tfrac{1}{6}\mathbf{t}_2 - \tfrac{1}{4}\mathbf{t}_3)\}. \tag{7.21}$$

The Brillouin zone for the h.c.p. structure is illustrated in fig. 21. If we write $\mathbf{K}$ as

$$\mathbf{K} = \alpha \mathbf{g}_1 + \beta \mathbf{g}_2 + \gamma \mathbf{g}_3 \tag{7.22}$$

Fig. 21

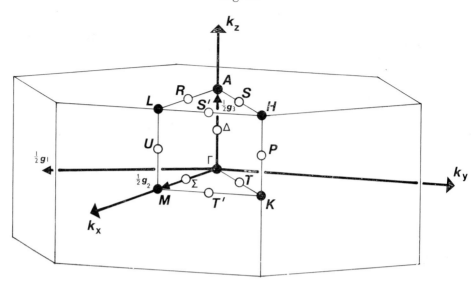

Brillouin zone for the h.c.p. structure.

eqn. (7.19) becomes

$$\hat{F}(\mathbf{K}) = 2 \cos \{2\pi(\tfrac{1}{6}\alpha - \tfrac{1}{6}\beta - \tfrac{1}{4}\gamma)\} \tag{7.23}$$

so that $\hat{F}(\mathbf{K})$ will vanish if $2\pi(\tfrac{1}{6}\alpha - \tfrac{1}{6}\beta - \tfrac{1}{4}\gamma)$ is $(2n+1)\pi/2$, that is if

$$\tfrac{1}{3}\alpha - \tfrac{1}{3}\beta - \tfrac{1}{2}\gamma = n + \tfrac{1}{2} \tag{7.24}$$

where $n = 0, \pm 1, \pm 2, \ldots.$ These are the equations of a set of planes and therefore the structure factor can be expected to vanish for wave vectors $\mathbf{K}$ which terminate on one of these planes. Values of α, β and γ for some wave vectors of special symmetry in the Brillouin zone are given in table **37** where the notation is that of Bradley and Cracknell (1972). By determining the values of $-\tfrac{1}{3}\alpha + \tfrac{1}{3}\beta + \tfrac{1}{2}\gamma$ at the points of symmetry it is clear that the plane

$$-\tfrac{1}{3}\alpha + \tfrac{1}{3}\beta + \tfrac{1}{2}\gamma = \tfrac{1}{2} \tag{7.25}$$

passes between Γ and H and intersects KH and LH, cutting KH at a point of tri-section and cutting LH at its midpoint. These intersections will, of

Table 37. Some wave vectors of special symmetry in the h.c.p. Brillouin zone

Label	α	β	γ
Γ	0	0	0
M	0	$\tfrac{1}{2}$	0
A	0	0	$\tfrac{1}{2}$
L	0	$\tfrac{1}{2}$	$\tfrac{1}{2}$
K	$-\tfrac{1}{3}$	$\tfrac{2}{3}$	0
H	$-\tfrac{1}{3}$	$\tfrac{2}{3}$	$\tfrac{1}{2}$
$P(KH)$	$-\tfrac{1}{3}$	$\tfrac{2}{3}$	γ
$S'(LH)$	-2α	$\tfrac{1}{2}+\alpha$	$\tfrac{1}{2}$

course, be repeated by symmetry all round the Brillouin zone. Thus we see that there is no selection rule preventing one-magnon inelastic scattering of thermal neutrons by a ferromagnetic h.c.p. metal for any **K** except for values of **K** terminating on the plane specified in eqn. (7.25) and generated from this by the hexagonal symmetry of the Brillouin zone. The vanishing of $\hat{F}(\mathbf{K})$ on these planes does not, therefore, constitute a serious barrier to the experimental determination of magnon frequencies in such metals, although we may expect the scattered intensities to be small for scattering vectors **K** which terminate in the vicinity of one of these planes.

7.4. *Electron scattering and tunnelling*

Having just devoted some attention to the group-theoretical aspects of the inelastic scattering of neutrons, it is natural that we should now turn to the consideration of the group-theoretical aspects of the scattering of beams of electrons, which are the other kind of particle that has been used extensively in the study of the structures and properties of solids. We have already noted some of the important practical distinctions between inelastic neutron scattering and inelastic electron scattering. The principal differences arise from the following physical features :

(i) electrons are electrically charged whereas neutrons are uncharged,

(ii) the absorption/scattering cross sections, at least for low energy electrons, are very much higher than the corresponding cross sections for neutrons.

The existence of the electric charges on the electrons means that Coulomb interactions provide an important mechanism contributing to the scattering for beams of electrons but not for beams of neutrons. The very small penetration depth for low-energy electrons means that it is likely to be surface phonons and surface magnons, rather than the corresponding bulk excitations, that are relevant to inelastic electron scattering. The small penetration depth also means that, for low-energy electrons, multiple scattering occurs ; if serious multiple scattering occurs this will prevent the determination of the energy or wave vector of an individual quasiparticle. There is some evidence that this problem may be able to be overcome by using high-energy electrons (Boersch *et al.* 1966, De Wames and Vredevoe 1967, Fujiwara and Ohtaka 1968) ; the difficulty then is of achieving sufficiently good monochromatization of the incident beam and sufficiently good energy resolution of the scattered beam, because the phonon or magnon energies are quite tiny (~ tens of meV) compared with the electron energies (~ tens of keV) (for discussion see Kuyatt and Simpson (1967)). The experimental development of electron diffraction as a technique for investigating crystalline solids was delayed for a long time by various technical problems. In addition to the monochromatization of the incident beam and the energy analysis of the scattered beam, these problems also involved the preparation and characterization of the surfaces of specimens as well. It is only within the last decade or so that LEED has become an important standard experimental technique. With the development of LEED (elastic scattering) techniques, principally for the study of the structures of surfaces, it has also become possible to develop related techniques for inelastic scattering with the intention of studying surface state's that is surface electronic states, surface vibrations, and surface spin waves. One of the main difficulties is to separate the inelastic scattering from the elastic scattering. The quantitative results that have been obtained so far from inelastic electron scattering are rather few. It is the inelastic scattering of electrons which is of interest to us in this section ; the elastic scattering was discussed briefly in § 1.2. Because of the small number of experimental results that have been obtained so far it is, perhaps, not surprising that there has so far been hardly any use of group theory in connection with the inelastic scattering of electrons. Therefore, in this subsection we are only able to indicate areas in which it seems that group theory may be useful, rather than areas in which group theory has already been proved to be successful.

In the consideration of inelastic electron scattering it is convenient to make a somewhat artificial separation between scattering that involves electronic states of a ' frozen lattice ' and scattering that involves the excitation, or de-excitation, of quasiparticles such as phonons or magnons. This represents an extension of the well-known Born–Oppenheimer approximation to include spin waves as well. In the Born–Oppenheimer approximation one studies the lattice vibrations of a crystal, ignoring the effect of any changes which are induced in the electronic band structure by the lattice vibrations. Similarly one calculates the electronic band structure assuming each atom is frozen in its mean position.

Suppose we consider an incident electron, with energy E_0 and momentum $\hbar \mathbf{k}_0$, which is scattered by the excitation of phonons or magnons into a scattered-beam state, with energy E_1 and momentum $\hbar \mathbf{k}_1$. If it is possible to ensure that this electron has only suffered one scattering event, then measurement of E_0, E_1, $\mathbf{k}_0$ and $\mathbf{k}_1$ would enable one to determine the energy, $\hbar \omega(\mathbf{k})$, and wave vector, $\mathbf{k}$, of the quasiparticle involved. This would then be analogous to the neutron-scattering case, see eqns. (7.11) and (7.12). It would then be possible to establish selection rules for this scattering by using an argument analogous to that used in § 7.3 for neutron scattering. The principal difference, of course, would be that instead of the cross sections for the scattering of neutrons by phonons or magnons (see eqn. (8) of Elliott and Thorpe (1967) and eqn. (9) of Cracknell and Sedaghat (1973)) one would have to use the corresponding cross sections for electron scattering. The other important difference is that one would need to distinguish between scattering by bulk excitations (phonons or magnons) and scattering by surface excitations (phonons or magnons). For scattering by surface excitations it would be necessary to use the group theory of two-dimensional space groups and two-dimensional Brillouin zones, which has been described in § 5.3, instead of the more conventional three-dimensional treatment that is used for the bulk excitations. The identification of Van Hove singularities for the surface states could be achieved by adapting the procedure described in § 6.6, for three-dimensional space groups, to the case of the two-dimensional space groups and two-dimensional Brillouin zones for the surface states.

It appears that the group-theoretical analysis of the cross sections for inelastic electron scattering has not yet been examined in detail. The arguments indicated in the previous paragraph for the determination of selection rules for the inelastic scattering of electrons, assume that each electron has experienced only one scattering event involving the creation or annihilation of a quasiparticle. However, we have already noted that, at least at low energies, an electron may have experienced more than one scattering event. If this occurs the equations obtained from the conservation of energy and the conservation of momentum will not give the values of $\hbar\omega(\mathbf{k})$ and $\mathbf{k}$ for individual quasiparticle states, but only the total energy and the total momentum of an arbitrary collection of quasiparticles with completely unrelated values of $\mathbf{k}$. Even if one of these scattering events involves a quasiparticle with a special wave vector, the remaining scattering events will almost certainly involve quasiparticles with general wave vectors ; therefore it is unlikely that any useful selection rules can be obtained for electrons which suffer more than one scattering event. Nevertheless, there is still a possible use for group theory in multiple scattering by simplifying the reduced transfer matrix (see below). The multiple scattering is not to be confused with the two-phonon or two-magnon absorption or scattering processes which were considered in § 7.2 ; in that section we were concerned with a single scattering event involving two quasiparticles, whereas here we are concerned with the possibility of a succession of two or more scattering events between the incidence and emergence of the beam. The possibility of any one of these events being a two-phonon or two-magnon process is, of course, not excluded.

We now turn to the consideration of the absorption of an incident electron beam by processes involving the (excited) electronic states of a crystal. The possibility of using beams of electrons to study the electronic states of a crystal was appreciated by Bethe (1928) very soon after the first electron scattering work of Davisson and Germer (1927) and Thomson and Reid (1927). Bethe's approach involved trying to relate the intensities of the scattered and incident electron beam via (excited) electronic Bloch states within the crystal and we have already seen in §§ 2.2, 6.1 and 6.2 that band structures, for bound electronic states in solids, provided one of the first fields for the application of group theory in solid state physics. The incident electrons may be regarded as entering excited (or ' hot ') states, which may be bulk or surface electronic states, in the crystal. Alternatively, they may be regarded as being responsible for the excitation of electrons already in the crystal, below the Fermi energy, into excited bulk or surface states. Since the electrons are indistinguishable from one another these are equivalent descriptions of the same process. Subsequently the electrons in excited states will be scattered back to states with lower energy. Many of the developments in the theory of the scattering of electrons by solids have, quite naturally, been expressed in terms of the language and ideas of electronic band structure. Many workers regard this as the most fruitful approach to electron scattering and consequently adaptations of some of the group-theoretical ideas involved in band structures may become relevant to electron scattering. However, other workers (McRae 1966, 1968, Kambe 1967, Duke and Tucker 1969, Jennings 1970, 1971) have felt that theories of electron scattering, principally at low energies, have been too closely tied to the concepts and notation of band theory and that this has obscured some of the important physical aspects of the scattering process. An alternative approach has therefore also been developed in which one calculates the reflection and transmission coefficients for a beam of electrons incident on a plane of atoms parallel to the crystal surface ; the scattering of the incident beam of electrons is then considered in terms of multiple scattering from the planes of atoms parallel to the surface. One rather useful feature of this multiple scattering approach is that it can very easily be adapted to the problem of electron scattering by a crystalline specimen with a thin coating of another material on its surface. Each of the two approaches that we have mentioned involves some group-theoretical aspects and we shall consider them both in turn.

The conventional theory of the electronic band structure of a crystalline solid is based on the assumption of a finite-sized specimen with Born–von Kármán cyclic boundary conditions imposed at the surfaces of the specimen. Thus, effectively, the Bloch wave functions which are obtained, and which are characterized by wave vectors $\mathbf{k}$, can be regarded as eigenfunctions of the Hamiltonian of a system which is infinite in three dimensions. In conventional discussions of band structures the wave vectors $\mathbf{k}$ and the energy eigenvalues are tacitly assumed to be real. This is because for a complex wave vector $\mathbf{k}$ the eigenfunctions will be damped waves, decreasing in amplitude in one direction but increasing in the opposite direction. In an infinite crystal the exponentially increasing feature of these functions makes them unacceptable physically. However, in an electron-scattering problem the specimen can be regarded as a semi-infinite medium, with the plane containing the surface on which the electrons impinge as the plane separating the occupied region of space from the unoccupied region of space outside the crystal. Those eigenfunctions of the Hamiltonian which have complex wave vectors then acquire a physical significance, because it is possible to select those eigenfunctions which are exponentially decreasing as one proceeds away from the boundary in a direction normal to the surface. These can be regarded as electronic surface states. The exponentially decaying waves are also used in describing electron tunnelling effects. The boundary conditions that apply to the scattering of electrons by a crystal are worth noting ; they are :

(i) conservation of total energy of the incident electron,
(ii) conservation of the tangential component of the momentum of the incident electron,
(iii) continuity of ψ and $\nabla\psi$ across the surface,
(iv) finite electron densities everywhere.

A more extensive discussion of the boundary conditions is given, for example, by Stern and Taub (1970). Our interest in this article, however, is primarily in group-theoretical problems in solid-state physics, and so we shall now consider the possibility that $E(\mathbf{k})$ and $\mathbf{k}$ may be complex and examine the symmetry of $E(\mathbf{k})$ under these conditions (see, for example, Dederichs (1972)).

For the scattering of electrons by a crystal the energy, E, is real and is determined by the energy of the incident electrons. Therefore, using the first boundary condition, once it has entered the crystal the electron must be in a state with energy E, whether this is a surface state, with complex $\mathbf{k}$, or a bulk state, with real $\mathbf{k}$. If the wave vector $\mathbf{k}$ is resolved into two components, k_x and k_y, in the plane of the surface and a third component, k_z, normal to the surface, then (k_x, k_y) is still a ' good quantum number ' and its properties are defined by the two-dimensional space group and the two-dimensional Brillouin zone of the crystal surface. It is only the k_z component of $\mathbf{k}$ that may be complex. We are then interested in the band structure $E_j(\mathbf{k})$, which is a complex function of the complex z component k_z, with (k_x, k_y) being given and real. In particular, we require those complex k_z values that have the particular real value of the energy $E_j(\mathbf{k}) = E$. These k_z values lie on lines in the two-dimensional k_z plane which are described as 'real lines', where it is $E_j(\mathbf{k})$ and not necessarily k_z, which is real. The symmetry of $E_j(\mathbf{k})$ for real $E_j(\mathbf{k})$ and real $\mathbf{k}$ has already been described in § 6.2 ; this can be summarized by saying that $E_j(\mathbf{k})$ has the symmetry of a certain space group $\mathbf{G'}$ that is related to the space group $\mathbf{G}$ of the crystal in a prescribed manner. The symmetry of $E_j(\mathbf{k})$ as a function of $\mathbf{k}$ was determined by studying the transformation properties of Bloch functions under space-group operations, see eqn. (6.17). To these symmetry properties of $E_j(\mathbf{k})$ we can add, for complex $E_j(\mathbf{k})$, the condition

$$E_j(\mathbf{k}) = E_{j'}{}^{*}(\mathbf{k}^{*}) \tag{7.26}$$

where $E_j(\mathbf{k})$ and $E_{j'}{}^{*}(\mathbf{k})$ generally, but not necessarily, refer to the same band ; eqn. (7.26) results from the reality of the potential $V(\mathbf{r})$. A considerable insight can be obtained by considering a 'one-dimensional' band structure (Heine 1963, 1964), while the three-dimensional case is considered on pages 149–53 of Dederichs (1972).

The alternative approach to the theory of electron scattering, namely the multiple scattering approach, is a generalization of a theory put forward a long time ago by Darwin (1914) for X-ray scattering ; the development of these ideas for electron scattering is due to McRae (1966, 1968) and Kambe (1967). We follow the treatment given by McRae (1968) in which spin is ignored ; the effect of the inclusion of spin has been considered by Jennings (1970). Darwin's theory included only two beams, an incident beam and one diffracted beam ; this may be a good approximation for the (elastic) scattering of X-rays, but for the (inelastic) scattering of electrons one has to allow for the incident beam of electrons to be scattered into all possible propagating crystal states (Bloch states) and all possible evanescent-wave states (surface states). In practice one can only include a finite number, N, of forward-scattered beams and of back-scattered beams. We assign amplitudes $a_\mu(\mathbf{k}_v{}^+)$ and $b_\mu(\mathbf{k}_v{}^-)$ to the components with wave vectors $\mathbf{k}_v$ in the forward-scattered and back-scattered beams, respectively, at the layer of atoms labelled by μ. Using the calculated values of the reflection coefficients and transmission coefficients for a single layer of atoms one can obtain a recurrence relation between $(a_{\mu+1}(\mathbf{k}_v{}^+), b_{\mu+1}(\mathbf{k}_v{}^-))$ $(v = 1, 2, ..., N)$ and $(a_\mu(\mathbf{k}_w{}^+), b_\mu(\mathbf{k}_w{}^-))$ $(w = 1, 2, ..., N)$. Regarding the set of $2N$ coefficients for the layer μ as a vector c_μ in $2N$ dimensions, this recurrence relation can be represented by a $2N$ by $2N$ square matrix Q_μ, which is called the 'transfer matrix' for the layer μ ; thus

$$c_{\mu+1} = Q_\mu c_\mu. \tag{7.27}$$

By a repeated use of this recurrence relation the coefficients for the layer μ can then be related to the coefficients for the outermost layer

$$c_\mu = \prod_{v=0}^{\mu-1} Q_v c_0 \quad (\mu > 0). \tag{7.28}$$

The generalized Darwin problem is then to solve eqn. (7.28) for b_0, which is the lower half of c_0 and which represents the amplitudes of the reflected waves at the outermost layer, in terms of a_0, which is the upper half of c_0 and represents the amplitude(s) of the incident wave(s).   After some manipulation and making some assumptions (for details see McRae (1968)) the transfer matrices Q_μ can be replaced by a single 'reduced transfer matrix', $\mathbf{R}$, which is the same for all layers.

The application of group theory in this problem arises in connection with the diagonalization of $\mathbf{R}$, which is a time-consuming part of any calculation based on the multiple-scattering approach.   It may be possible to use symmetry arguments to get $\mathbf{R}$ into a simplified block-diagonal form.   If the Bloch states which are involved within the crystal belong to wave vectors on special lines of symmetry in the Brillouin zone, we can say that scattering will only occur between Bloch states which have the same symmetry, that is which belong to the same irreducible representation of the space group $\mathbf{G}$.   Thus, for example, for an electron beam incident normally on a (001) face of a single crystal of tungsten, which has the b.c.c. structure, the conservation of the tangential components of the electrons' momenta means that for the bulk states only those electronic states with $\mathbf{k}$ along the line of symmetry Δ ([001]) in the Brillouin zone will be involved.   The Bloch states involved in the scattering in this case will therefore belong to the irreducible representations Δ_1, Δ_1', Δ_2, Δ_2' or Δ_5 of $Im3m$ (O_h^9), where Δ_5 is two-fold degenerate.   For surface states the conservation of the tangential components of the electrons' momenta means that only surface states at the centre point, Γ, of the two-dimensional Brillouin zone of the surface can be involved.   If we consider Bloch states along Δ derived from s, p and d atomic wave functions they will belong to the following representations (Sedaghat and Cracknell 1974)

$$s : \Delta_1$$
$$p : \Delta_1, \Delta_5$$
$$d : \Delta_1, \Delta_2, \Delta_2', \Delta_5.$$

The number of states belonging to each representation in this case is therefore

Δ_1	3
Δ_1'	0
Δ_2	1
Δ_2'	1
Δ_5	4 (two two-fold degenerate states)

and the value of N is 9.   The 18 by 18 reduced transfer matrix $\mathbf{R}$ can then be expressed in a simplified block-diagonal form.   There will be four blocks, corresponding to Δ_1, Δ_2, Δ_2' and Δ_5 states.   For example, the Δ_1 states will be able to scatter only into Δ_1 states and not into Δ_2, Δ_2' and Δ_5 states ; there are three Δ_1 states involved and so they lead to a 6 ($= 2 \times 3$) by 6 block in the reduced transfer matrix, $\mathbf{R}$.   Arguing in a similar manner for the other representations one finds that the four blocks of non-zero elements down the leading diagonal of the block-diagonalized form of $\mathbf{R}$ have the dimensions 6 by 6 (Δ_1), 2 by 2 (Δ_2), 2 by 2 (Δ_2') and 8 by 8 (Δ_5).   This simplification of the form of $\mathbf{R}$ considerably reduces the computational effort required to diagonalize the matrix $\mathbf{R}$.

We should also consider what happens if the incident electron beam departs from normal incidence. For a few special values of the orientation and energy of the incident electron beam, it may happen that the conservation of the tangential components of the electrons' momenta leads to the excitation of bulk states or surface states with wave vectors with some special symmetry. In this case some group-theoretical simplification of the form of **R** will still be possible. However, for arbitrary orientations and energies of the incident electron beam, the wave vectors of the Bloch states that are involved in the scattering will not be wave vectors with any special symmetry ; therefore in general no group-theoretical simplification of the reduced transfer matrix **R** will be possible.

§ 8. Phase transitions

At the outset it is convenient to distinguish between two different kinds of phase transitions in solids. We do not refer to the thermodynamic classification introduced by Ehrenfest and based on the continuity properties of the Gibbs function, G, and its derivatives. Instead we refer to the distinction between (i) phase transitions which involve a drastic structural re-arrangement of the relative positions of the atoms in the solid and (ii) phase transitions which do not involve such a structural re-arrangement. An example of (i) would be a phase transition of a simple metal from, say, one of the three common metallic structures (b.c.c., f.c.c., and h.c.p.) to another of these structures ; thus

$$\text{Li and Na :} \quad \text{b.c.c.} \rightarrow \text{h.c.p. at low temperatures,}$$

$$\left.\begin{array}{l} \text{Ca :} \quad \text{f.c.c.} \rightarrow \text{h.c.p.} \\ \text{Sr :} \quad \text{f.c.c.} \rightarrow \text{b.c.c.} \end{array}\right\} \text{at high temperatures.}$$

In (ii) there is no significant change in the positions of the atoms in the crystal and the phase transition only involves changes in the values of some tensor property, A, that is associated with the atoms. A may be a magnetic dipole moment or an electric dipole moment. For example, if a single crystal of MnF_2 is cooled from a temperature T_1, which is very much higher than $\sim 67 \cdot 3$ K, to a temperature T_2, which is very much lower than $\sim 67 \cdot 3$ K, the crystal will undergo a transition from the paramagnetic phase to the antiferromagnetic phase. In the high-temperature phase the spins of the Mn atoms (or ions) are randomly oriented relative to one another and relative to the crystal axes. In the low-temperature phase the magnetic moments of the Mn atoms are antiferromagnetically ordered in the manner illustrated in fig. 1. There is also the possibility of a magnetic phase transition that occurs between two different magnetically ordered phases without any drastic structural re-arrangement of the atomic positions. Another example of (ii) would be a ferroelectric transition that was not accompanied by a drastic structural re-arrangement of the atomic positions. When we speak of 'drastic structural re-arrangement' we do not, of course, mean to use this term to include the magnetostrictive or 'electrostrictive' (piezoelectric) distortions which are frequently found to be associated with such transitions. These distortions are, usually, quite small and can easily be distinguished from 'drastic structural re-arrangements' of crystals. For example, in the case of anti-ferromagnetic NiF_2 (see fig. 2) the lattice constants a and b, which are equal in the tetragonal paramagnetic phase, are

$$a = (4 \cdot 648 \ 44 \pm 0 \cdot 000 \ 04) \ \text{Å}$$

$$b = (4 \cdot 647 \ 19 \pm 0 \cdot 000 \ 04) \ \text{Å}$$

(Haefner *et al.* 1966) and the magnetostrictive distortion can be seen to be extremely small. It is difficult to place a quantitative demarcation between the magnitudes of the changes in the relative positions of the atoms in drastic structural re-arrangements and in magnetostrictive, or other similar, distortions.

Group theory cannot be used to give very much useful information about phase transitions that involve drastic structural re-arrangements because there is no useful relationship between the space groups $\mathbf{G}_1$ and $\mathbf{G}_2$ of the two phases involved. The theoretical prediction of which of these two structures is actually stable at any given temperature would involve the calculation, presumably from first principles, of the total energy of the material, at the chosen temperature, for each of these two structures. The structure with the lower energy would then be the stable one at that temperature ; the phase transition corresponds to a crossing of the curves of energy versus temperature for these two structures. On the other hand, if a phase transition occurs without any drastic structural re-arrangement, there will be some useful relationship between the space groups $\mathbf{G}_1$ and $\mathbf{G}_2$ of the material in the two phases. In this section we shall be concerned with various aspects of the exploitation of this relationship between $\mathbf{G}_1$ and $\mathbf{G}_2$ and our remarks will, therefore, generally be restricted to phase transitions which do not involve a drastic structural re-arrangement of the atomic positions.

8.1. *Structure determinations, 'forbidden' effects, and representation analysis*

From comments that we have already made in various sections of this article it should be apparent that standard, and fairly routine, methods exist for the determination of crystal structures. The relative positions of the atoms in a crystal, and therefore the classical space group to which the crystal belongs, are most commonly determined from the analysis of the results of X-ray diffraction experiments. However, although X-ray diffraction provides a very powerful technique for the determination of the structure of any given crystal, it has always been good practice among mineralogists or crystallographers to advocate the limitation of the number of possible structures of a given crystal by a few simple arguments, before resorting to the use of X-ray diffraction.

The point-group symmetry of a crystal can often be determined uniquely, or reduced to one of a small number of possibilities, by quite simple methods. This in turn reduces the number of possible space groups. The practical methods for the determination of point group symmetry are described in standard textbooks on crystallography and do not need to be discussed here in detail (see, for example, chapter 11 of the book by Buerger (1963)). Apart from the obvious, but slightly tedious, technique of optical goniometry and possibly the study of etch figures and dissolution forms, the methods available involve the exploitation of the theory of equilibrium tensor properties which we have already mentioned in § 3.1. These methods include the examination of the crystal with an optical polarizing microscope and the testing of the crystal to see whether it exhibits optical activity, pyroelectricity, ferroelectricity, or piezoelectricity. In the polarizing microscope one is studying the form of the refractive index which is related to the permittivity tensor ; the form of the permittivity is, in turn, related to the point group symmetry of the crystal, see table 6. The pyroelectric, or ferroelectric, and piezoelectric tensors have already been identified in eqns. (3.8) and (3.9). If pyroelectricity, ferroelectricity, or piezoelectricity can be detected experimentally for a given material, this eliminates the possibility of a number of high-symmetry point groups, see table 7 for example. The failure to detect any one of these phenomena experimentally does not necessarily, however, prove the existence of one of those point groups in which the appropriate phenomenon is ' forbidden '—the effect may be present but too small to detect with the apparatus available. Extensive further discussions of symmetry and the tensor properties of crystals will be found, for example, in the books by Nye (1957) and Wooster (1973).

For a crystal which is magnetically ordered, one could attempt a complete determination of its structure by using neutron-diffraction techniques. However, the use of neutron diffraction is cumbersome and expensive and it is prudent to obtain as much information as possible previously by other means. The positions of the atoms in the crystal can be determined from X-ray diffraction experiments. If the transition to the magnetically ordered state, in a given magnetic crystal, involves only the ordering of the magnetic moments, without any changes in the positions of the atoms, the X-ray data can be obtained from work performed at temperatures either above or below the temperature at which the transition to the magnetically ordered phase occurs. For many crystals in which magnetic ordering has been suspected or detected, the positions of the atoms were already known and so could simply be obtained from one of the published lists of crystal structures. Assuming that the positions of the atoms are known, it remains to find the orientations of the ordered magnetic moments relative to the crystallographic axes. For a simple ferromagnetic crystal this information can be obtained from magnetization measurements on single-crystal specimens. For crystals with more complicated magnetic ordering patterns other methods have to be used. Apart from the use of elastic neutron scattering, which provides by far the most important method for the determination of the orientations of the magnetic moments in a magnetically ordered crystal, there are several other methods which have been suggested and exploited. They are similar in their philosophy to the methods using ' forbidden ' effects etc. which we mentioned in connection with non-magnetic crystals. The study of the form of the magnetoelectric tensor (see table 9) has been particularly successful in this connection (see, for example, Hornreich (1972) and also § 2.1 of Cracknell (1974 a)) ; indeed it has enabled some magnetic structures to be determined uniquely without the use of neutron diffraction at all. Other methods which have been suggested involve the use of magnetic susceptibility measurements, nuclear magnetic resonance, the Mössbauer effect, and the study of piezo-magnetism, ferroelectricity, and Davydov splitting.

In addition to the problem of the determination of magnetic structures there is also the problem of the description of the symmetries of magnetic structures in terms of groups of symmetry operations. In § 1.2 we introduced generalized space-group operations denoted by $\{R_i{}^A | R_i{}^r | \mathbf{v}_i\}$, where the rotations $R_i{}^r$ for the position vector of an atom and $R_i{}^A$ for the tensor property need not necessarily be the same operation. However, we subsequently introduced the restriction that $R_i{}^r$ and $R_i{}^A$ should be the same rotational operation and the tensor $A(\mathbf{r})$ was taken to be an axial vector, or pseudo-vector, namely the spin, or magnetic moment, of the atom at $\mathbf{r}$. Thereby we obtained the Heesch–Shubnikov point groups and space groups. In those succeeding sections in which we have been involved with magnetically ordered crystals we have always used the Heesch–Shubnikov groups to describe the symmetries of these crystals. However, the examples which we have considered have usually been crystals exhibiting a rather simple form of magnetic ordering, such as simple ferromagnetic metals and simple antiferromagnetic insulators such as MnF_2. In these cases the use of the Heesch–Shubnikov groups is quite successful. But the fact that these groups are suitable for simple examples does not necessarily mean that they are particularly suitable for crystals with more complicated magnetic-ordering patterns. The condition that we only include operations in which we have the same operation acting on both the position vectors and on the magnetic moment vectors is often unnecessarily restrictive. This is particularly so when the period of the magnetic ordering pattern is incommensurate with the periodicity of the Bravais lattice of the space group, $\mathbf{G}$, that describes the symmetry of the positions of the atoms when magnetic ordering is neglected. Consequently an alternative approach to the description of magnetic structures has been developed, principally by Bertaut (see, for example, Bertaut (1968 a, 1971)). This approach is called 'representation analysis' and is based on the use of the irreducible representations of the classical space group, $\mathbf{G}$, which describes the positions of the atoms, neglecting their magnetic moments. However, representation analysis is much more powerful than just providing a scheme for the labelling or classification of magnetic structures. It is concerned with the interpretation of experimental results in the search for the structure, before it is known, and also with the discussion of the interactions which might explain the occurrence of the actual structure that is finally deduced.

The analysis of magnetic structures by representation analysis is based on the transformation properties of the magnetic moments, commonly referred to as 'spins' $\mathbf{S}_p$, of the atoms or ions, under the operations of the classical space group, $\mathbf{G}$, of the atomic positions or under the operations of the sub-group $\mathbf{G}^k$. The transformation of the $\mathbf{S}_p$ will generate a representation, which we may call $\Gamma_S{}^k$,

$$\{R|\mathbf{v}\}\mathbf{S}_p = \sum_q \mathbf{D}_S{}^k(\{R|\mathbf{v}\})_{pq}\mathbf{S}_q. \tag{8.1}$$

This is analogous to the generation of the representation $\Gamma_\xi{}^k$ from the transformation properties of the displacements of the atoms or ions in a crystal when discussing lattice vibrations in § 6.3. The differences are first that the spins are axial vectors, or pseudo-vectors, instead of polar vectors, and secondly that we are describing a static spin configuration and not a dynamical situation. If the operation $\{R|\mathbf{v}\}$ moves an atom to a position outside the chosen unit cell it can be related to one of the atoms in the unit cell by a suitable phase factor, see eqns. (6.36)–(6.38).

The choice of $\mathbf{k}$ is governed by the relationship between the magnetic unit cell and the 'chemical' unit cell, that is the unit cell of $\mathbf{G}$:

    (i) magnetic unit cell = chemical unit cell ; $\mathbf{k} = \mathbf{0}$,

    (ii) magnetic unit cell = some simple multiple of chemical unit cell ; $\mathbf{k}$ is some rational fraction of a reciprocal lattice vector of $\mathbf{G}$ and is equal to a reciprocal lattice vector for the lattice of the magnetic structure,

    (iii) magnetic unit cell is infinite, i.e. magnetic ordering is incommensurate with the periodicity of the lattice of $\mathbf{G}$; $\mathbf{k}$ = wave vector of magnetic ordering pattern.

Of course, if the magnetic structure is unknown, $\mathbf{k}$ has to be determined from experimental results. Thus if all the observed lines in a neutron-diffraction pattern can be indexed in terms of the chemical unit cell it means that the magnetic unit cell is the same as the chemical unit cell and therefore $\mathbf{k} = \mathbf{0}$. In the case of antiferromagnetic MnO, which was the first magnetic structure to be determined successfully from neutron diffraction work, lines were observed which could only be indexed in terms of an enlarged unit cell that implied the existence of a wave vector $\mathbf{k} = (2\pi/a)(\frac{1}{2}, \frac{1}{2}, \frac{1}{2})$ so that

$$\exp(i\mathbf{k} \cdot \mathbf{r}) = \exp\{\pi i(u+v+w)\} = \sigma(\mathbf{r}) \tag{8.2}$$

where the coordinates of the Mn atoms are $(u, v, w)a$ and a is the length of the side of the chemical unit cell. It is easy to check that $\sigma(\mathbf{r})$ gives the correct signs for the spins of the Mn atoms, see fig. 22. Thus for $(2, 0, 0)a$, $(0, 2, 0)a$, $(1, 1, 0)a$, etc. $\sigma(\mathbf{r}) = +1$, while for $(1, 0, 0)a$, $(0, 1, 0)a$, $(\frac{1}{2}, \frac{1}{2}, 0)a$, $(1, 1, 1)a$, etc. $\sigma(\mathbf{r}) = -1$. $\sigma(\mathbf{r})$ takes the same value throughout any given plane defined by $\mathbf{k} \cdot \mathbf{r} = $ constant, that is throughout any plane normal to $\mathbf{k}$.

Fig. 22

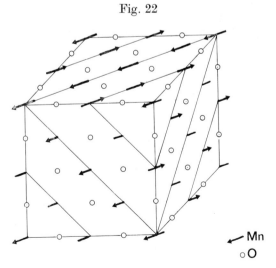

Unit cell of antiferromagnetic MnO (Roth 1960).

This example also illustrates the analogy between Bloch waves and the static spin arrangement. If $\mathbf{S}(\mathbf{r})$ is the spin of the atom at $\mathbf{r}$, then the spin of an atom at $\mathbf{r} + \mathbf{T}$, where $\mathbf{T}$ is a lattice translation, is

$$\mathbf{S}(\mathbf{r} + \mathbf{T}) = \exp(i\mathbf{k} \cdot \mathbf{T})\mathbf{S}(\mathbf{r}). \tag{8.3}$$

In the example of MnO the function $\exp(i\mathbf{k} \cdot \mathbf{r})$ takes one or other of the two values $+1$ and -1. There is an obvious generalization for helical and sinusoidal spin configurations. We write

$$\mathbf{S}(\mathbf{r}) = \mathbf{S}_\mathbf{k}(\mathbf{r}) + \mathbf{S}_{-\mathbf{k}}(\mathbf{r}) \tag{8.4}$$

where

$$\mathbf{S}_\mathbf{k}(\mathbf{r}) = e^{i\mathbf{k} \cdot \mathbf{r}}(\mathbf{i}_1 a\, e^{i\alpha} + \mathbf{i}_2 b\, e^{i\beta} + \mathbf{i}_3 c\, e^{i\gamma}) \tag{8.5}$$

and $\mathbf{S}_\mathbf{k}{}^*(\mathbf{r}) = \mathbf{S}_{-\mathbf{k}}(\mathbf{r})$. $\mathbf{i}_1$, $\mathbf{i}_2$ and $\mathbf{i}_3$ are unit vectors along the directions of the crystal axes, a, b and c are the lattice constants and α, β and γ are phase angles to be determined by experiment. $\mathbf{S}_\mathbf{k}(\mathbf{r})$ and $\mathbf{S}_{-\mathbf{k}}(\mathbf{r})$ behave like Bloch waves. In the case of helical or sinusoidal spin structures the neutron diffraction crystallographer determines $\mathbf{k}$ from satellite reflections.

Let us assume that the representation $\Gamma_S{}^\mathbf{k}$ has been constructed (see eqn. (8.1)). $\Gamma_S{}^\mathbf{k}$ may be reducible. The irreducible representations of $\mathbf{G}^\mathbf{k}$ are available in the standard sets of tables and therefore it is quite straightforward to determine the reduction of $\Gamma_S{}^\mathbf{k}$. Having found the irreducible representations of $\mathbf{G}^\mathbf{k}$ that are obtained on the reduction of $\Gamma_S{}^\mathbf{k}$ it is then possible to determine basis vectors for each of these representations and thereby to identify possible spin configurations. This can be done using projection operator techniques. If ψ is an arbitrary function we generate the function

$$\psi_{j,\,pq}{}^\mathbf{k} = \sum_{\{R|\mathbf{v}\} \in \mathbf{G}} \mathbf{D}_j{}^\mathbf{k}(\{R|\mathbf{v}\})_{pq}{}^* \{R|\mathbf{v}\}\psi. \tag{8.6}$$

$\mathbf{D}_j{}^\mathbf{k}(\{R|\mathbf{v}\})_{pq}$ is the p, q element of the matrix representative of $\{R|\mathbf{v}\}$ in the representation $\Gamma_j{}^\mathbf{k}$ of $\mathbf{G}$. The function $\psi_{j,\,pq}{}^\mathbf{k}$ generated by eqn. (8.6) then forms a basis of the irreducible representation $\Gamma_j{}^\mathbf{k}$ of $\mathbf{G}$. The right-hand side of eqn. (8.6) may simply vanish; this indicates that the original choice of ψ was of an unsuitable form.

Let us consider the example of the space group $Pbnm$ ($D_{2h}{}^{16}$) (Bertaut 1968 a) which is the classical space group $\mathbf{G}$ of the atomic positions in a number of rare-earth perovskite-type crystals, such as $ErFeO_3$ (Koehler et al. 1960), $ErCrO_3$ (Bertaut and Maréschal 1967), and $TbFeO_3$ (Bertaut et al. 1967). If the magnetic unit cell is identical with the chemical unit cell, we take $\mathbf{k} = \mathbf{0}$ and the irreducible representations of $\mathbf{G}^\mathbf{k}$ are then related in a trivial manner to the irreducible representations of the point group mmm (D_{2h}), see table 38. In these materials the transition metal atoms may be on the sites labelled a or b in the Wyckoff notation and the rare earth atoms are on the c sites; the coordinates of these sites are

$$
\left.
\begin{array}{llllll}
a & (1)\ 000 & (2)\ 00\tfrac{1}{2} & (3)\ \tfrac{1}{2}\tfrac{1}{2}0 & & (4)\ \tfrac{1}{2}\tfrac{1}{2}\tfrac{1}{2} \\[4pt]
b & (1)\ \tfrac{1}{2}00 & (2)\ \tfrac{1}{2}0\tfrac{1}{2} & (3)\ 0\tfrac{1}{2}0 & & (4)\ 0\tfrac{1}{2}\tfrac{1}{2} \\[4pt]
c & (1)\ xy\tfrac{1}{4} & (2)\ \bar{x}\bar{y}\tfrac{3}{4} & (3)\ \tfrac{1}{2}+x,\ \tfrac{1}{2}-y,\ \tfrac{3}{4} & (4)\ \tfrac{1}{2}-x,\ \tfrac{1}{2}+y,\ \tfrac{1}{4}
\end{array}
\right\}. \tag{8.7}
$$

Table 38. Irreducible representations and bases for $Pbnm$ at $\mathbf{k}=0$

	$\{C_{2x}\|\tfrac{1}{2}\tfrac{1}{2}\tfrac{1}{2}\}$	$\{C_{2y}\|\tfrac{1}{2}00\}$	$\{I\|000\}$	Spins on sites a or b			Spins on site c			Related Heesch–Shubnikov group	
											or
Γ_1^+	1	1	1	A_x	G_y	C_z	.	.	C_z	$Pbnm$	$Pnma$
Γ_2^+	1	-1	1	F_x	C_y	G_z	F_x	C_y	.	$Pbn'm'$	$Pn'm'a$
Γ_3^+	1	1	1	C_x	F_y	A_z	C_x	F_y	.	$Pb'nm'$	$Pnm'a'$
Γ_4^+	1	-1	1	G_x	A_y	F_z	.	.	F_z	$Pb'n'm$	$Pn'ma'$
Γ_1^-	1	1	-1	.	.	.	G_x	A_y	.	$Pb'n'm'$	$Pn'm'a'$
Γ_2^-	1	-1	-1	.	.	.	.	.	A_z	$Pb'nm$	$Pnma'$
Γ_3^-	1	1	-1	.	.	.	.	.	G_z	$Pbn'm$	$Pn'ma$
Γ_4^-	1	-1	-1	.	.	.	A_x	G_y	.	$Pbnm'$	$Pnm'a$

The number in brackets is the label p that is used to discriminate between equivalent positions. By studying the transformation properties of spin vectors $\mathbf{S}_p$ on any one set of equivalent sites, it is possible to construct $\Gamma_S{}^{\mathbf{k}}$ for that set of sites. Here each $\Gamma_S{}^{\mathbf{k}}$ will be a twelve-fold degenerate representation. $\Gamma_S{}^{\mathbf{k}}$ can then be reduced by inspection or by the use of the standard formula

$$a_j{}^{\mathbf{k}} = (1/|\mathbf{G}|) \sum_{G \in \mathbf{G}} \chi_S{}^{\mathbf{k}}(G) \chi_j{}^{\mathbf{k}}(G)^* \tag{8.8}$$

where $a_j{}^{\mathbf{k}}$ is the number of times that $\Gamma_j{}^{\mathbf{k}}$ appears in the reduction of $\Gamma_S{}^{\mathbf{k}}$. The details for this space group are given by Bertaut (1968 a, 1971). For the spins on the a or b sites we obtain

$$\Gamma_S{}^{\mathbf{k}} = 3\Gamma_1{}^+ \oplus 3\Gamma_2{}^+ \oplus 3\Gamma_3{}^+ \oplus 3\Gamma_4{}^+ \tag{8.9}$$

and for the c sites we obtain

$$\Gamma_S{}^{\mathbf{k}} = \Gamma_1{}^+ \oplus 2\Gamma_2{}^+ \oplus 2\Gamma_3{}^+ \oplus \Gamma_4{}^+ \oplus 2\Gamma_1{}^- \oplus \Gamma_2{}^- \oplus \Gamma_3{}^- \oplus 2\Gamma_4{}^-. \tag{8.10}$$

Basis vectors for these representations, in terms of spin components on the a, b or c sites are given in table 38. In table 38 the symbols A_m, C_m, F_m and G_m ($m = x, y, z$) have the following meanings:

$$\left.\begin{aligned}
A_x &= S_{1x} - S_{2x} - S_{3x} + S_{4x} \\[4pt]
C_x &= S_{1x} + S_{2x} - S_{3x} - S_{4x} \\[4pt]
F_x &= S_{1x} + S_{2x} + S_{3x} + S_{4x} \\[4pt]
G_x &= S_{1x} - S_{2x} + S_{3x} - S_{4x}
\end{aligned}\right\} \tag{8.11}$$

and similarly for $m = y$ and $m = z$. The label p in S_{pm} is given in the identification of the sites in (8.7). The physical significance of these quantities can be determined by inspection. For example F_x has a maximum value if $S_{1x} = S_{2x} = S_{3x} = S_{4x}$ and is zero for any antiferromagnetic sign combination; it characterizes a ferromagnetic configuration. A_x, C_x and G_x characterize various antiferromagnetic configurations. In $ErFeO_3$ and $ErCrO_3$ it was found that, at low temperatures, the spin components of the Fe or Cr belong to a G_m mode, with both G_x and G_y components, while those of the Er belong to a C_z mode. G_x belongs to $\Gamma_4{}^+$ and G_y and C_z belong to $\Gamma_1{}^+$. It is interesting to note that the G_x mode would be described by the Heesch–Shubnikov group $Pb'n'm$ but the G_y and C_z modes would be described by $Pbnm$ (see the final column of table 38). Thus the 'global' magnetic symmetry would be the intersection of these two magnetic groups, namely the monoclinic space group $P2_1/m$. This illustrates a point that we mentioned at the beginning of our discussion of representation analysis, namely that the definition of a symmetry operation for a Heesch–Shubnikov space group may be unnecessarily restrictive.

Let us suppose that the possible spin configurations for a given magnetic crystal have been determined. We now turn to the problem of constructing possible terms for inclusion in the spin Hamiltonian, $\mathcal{H}$, of a magnetic crystal. In the representation analysis approach it is assumed that the spin Hamiltonian, $\mathcal{H}$, is invariant under all the operations of $\mathbf{G}$, the space group which describes the symmetry of the configuration of the equilibrium positions of the atoms in the magnetically ordered crystal. We also impose the condition that the Hamiltonian is invariant under the reversal of all the spins. This is equivalent to saying that $\mathscr{H}$ is invariant under all the operations of the grey space group $\mathbf{G} + \theta\mathbf{G}$. The spin Hamiltonian must take the form

$$\mathscr{H} = \sum_{\substack{\mathbf{r}, \, \mathbf{r}' \\ m, \, m'}} A_{mm'}(\mathbf{r}, \, \mathbf{r}')\mathbf{S}_m(\mathbf{r})\mathbf{S}_{m'}(\mathbf{r}') + H_4 + H_6 + \dots \qquad (8.12)$$

where m, $m' = x, y, z$. $\mathbf{S}_m(\mathbf{r})$ is the m component of the spin of the atom situated at $\mathbf{r}$ and $A_{mm'}(\mathbf{r}, \mathbf{r}')$ is a 3 by 3 matrix which represents a tensor of rank 2. H_4 and H_6 represent terms of order 4 and 6, respectively, in the spins ; these terms will be neglected. One could use projection operator techniques to perform the required symmetrization of $\mathscr{H}$ in eqn. (8.12), using of course only the identity representation Γ_1 (or Γ_1^+) of $\mathbf{G}$. However, this may be quite cumbersome. It may be more convenient to resolve the individual spins in terms of the basis vectors. For example, from (8.11)

$$S_{1x} = A_x + C_x + F_x + G_x. \qquad (8.13)$$

It is well known that the product $\Gamma_i \boxtimes \Gamma_j$ of two (irreducible) representations of a group $\mathbf{G}$ contains the identity representation Γ_1 (or Γ_1^+) of $\mathbf{G}$ if Γ_i and Γ_j are equivalent representations, or if $\Gamma_j \equiv \Gamma_i^*$ for complex conjugate pairs of representations. Therefore to construct functions which are of order two in the spins and which are invariant under the operations of the group $\mathbf{G}$, we form products of two basis vectors that belong to the same irreducible representation $\Gamma_i^{\mathbf{k}}$ of $\mathbf{G}^{\mathbf{k}}$, or to $\Gamma_i^{\mathbf{k}}$ and $\Gamma_i^{\mathbf{k}*}$ if $\Gamma_i^{\mathbf{k}}$ is complex. Strictly speaking, these basis vectors belong to the induced representations $(\Gamma_i^{\mathbf{k}} \uparrow \mathbf{G})$ so that the product of the two bases really belongs to the identity representation of $\mathbf{G}$ rather than of $\mathbf{G}^{\mathbf{k}}$. If the representation $\Gamma_i^{\mathbf{k}}$ of $\mathbf{G}$ that is used to describe a given spin arrangement is non-degenerate, then Γ_1 (or Γ_1^+) is the only representation that is obtained in the reduction of $\Gamma_i^{\mathbf{k}} \boxtimes \Gamma_i^{\mathbf{k}}$ (or of $\Gamma_i^{\mathbf{k}} \boxtimes \Gamma_i^{\mathbf{k}*}$ if $\Gamma_i^{\mathbf{k}}$ is complex). Therefore, all the terms obtained from the product of two basis vectors in this manner have the full symmetry of the space group $\mathbf{G}$; they will all be suitable functions for inclusion in the spin Hamiltonian, $\mathscr{H}$, in eqn. (8.12). However, $\Gamma_i^{\mathbf{k}}$ may be a degenerate irreducible representation of $\mathbf{G}$; this occurs, for example, in the cubic material Mn_3GaN (see Bertaut 1971). In this situation the reduction of $\Gamma_i^{\mathbf{k}} \boxtimes \Gamma_i^{\mathbf{k}}$ (or $\Gamma_i^{\mathbf{k}} \boxtimes \Gamma_i^{\mathbf{k}*}$ if $\Gamma_i^{\mathbf{k}}$ is complex) will contain not only Γ_1 (or Γ_1^+) but also several other irreducible representations of $\mathbf{G}$ as well. In this case, therefore, it will be necessary to examine each of the terms appearing in the product of the basis vectors to determine whether that term belongs to Γ_1^+ or not ; this could be done either by inspection or by using projection operator techniques. If $\Gamma_i^{\mathbf{k}}$ is isomorphous with a point-group representation the distribution of these terms among the various representations obtained from the reduction of $\Gamma_i^{\mathbf{k}} \boxtimes \Gamma_i^{\mathbf{k}}$ can be obtained from the tables of coupling coefficients given by Koster *et al.* (1963) (see also § 1.5). Having described the procedure for symmetrizing $\mathscr{H}$ in eqn. (8.12) according to the space group $\mathbf{G}$, we should note that this is more restrictive than imposing the condition that $\mathscr{H}$ be invariant under the operations of the appropriate Heesch–Shubnikov space group which describes the symmetry of the magnetically ordered crystal.

This may be justified, in practice, on the grounds that there is not enough experimental data available to enable one to determine all the parameters that would be needed in a spin Hamiltonian which possessed only the symmetry of the Heesch–Shubnikov group.

The use of representation analysis is not restricted to the discussion of magnetically ordered crystals. It can also be applied to ferroelectric crystals (Bertaut 1968 b) (see also § 8.2). It can also be adapted to the discussion of magnetoelectric effects ; this involves products of basis vectors which are axial vectors for magnetic moments or magnetic fields, but polar vectors for electric dipole moments or electric fields.

8.2. *Ferroelectric transitions ; ferroelasticity*

In § 8.1 we noted that the procedure involved in the construction of $\Gamma_s{}^k$ for the static configuration of spins in a magnetically ordered crystal was analogous to the procedure involved in the construction of $\Gamma_z{}^k$ for the vibrations of the atoms in a crystal. The analogy is more than just a formality ; it provides a very illuminating way to look at ferroelectric phase transitions. Suppose that we consider a material which is not ferroelectric at high temperatures but becomes ferroelectric at low temperatures. The point group of the high-temperature phase may be one of those point groups in which a spontaneous electric dipole moment (polar vector, **P**) is forbidden, but in its low-temperature phase its symmetry must belong to one of those point groups in which a non-zero polar vector, **P**, is allowed. An example of such a crystal

Fig. 23

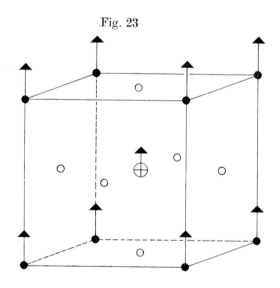

●   Ba or Sr

◯   Ti

○   O

Unit cell of $BaTiO_3$ or $SrTiO_3$. The arrows indicate displacements which occur in the ferroelectric phase.

would be $BaTiO_3$ or $SrTiO_3$, which have the perovskite structure at high temperatures, see fig. 23. This structure has cubic symmetry. The property of ferroelectricity in these materials is associated with displacements of the Ti and O atoms, relative to the Ba or Sr atoms, in a direction parallel to one of the edges of the cube ; these displacements are indicated by arrows in fig. 23. It is the relative displacements of the ions which give rise to the existence of a spontaneous electric polarization of the crystal, i.e. the property of ferroelectricity. These displacements of the atoms reduce the symmetry of the crystal from cubic to tetragonal and the point group of the symmetry of the crystal must, therefore, be one of those tetragonal point groups which allows the existence of a spontaneous electric dipole moment (polar vector, P), see table 7.

The displacements of the atoms which occur in a ferroelectric transition are, of course, invariant under the operations of the space group G_2 of the ferroelectric phase. But in studying a ferroelectric crystal one could also study these static displacements of the atoms, from their positions in the high symmetry phase, in terms of irreducible representations of G_1, the space group of the high symmetry, or paraelectric, phase. This is directly analogous to the use of representation analysis for magnetic structures which we described in § 8.1. It is therefore also analogous to the resolution of the lattice vibrations of a crystal with the symmetry of the space group G_1 in terms of normal modes of vibration belonging to the irreducible representations of G_1. Thus the displacements that are associated with the existence of the ferroelectric phase can be resolved in terms of the irreducible representations of G_1. The difference, of course, from the lattice vibrations case is that we are now concerned with a set of static displacements and not dynamic displacements. It came to be realized just before 1960 that these displacements could be regarded simply as the 'freezing in' of the displacements in one of the normal modes of the crystal (with space group G_1) as the crystal was cooled through the transition temperature. In other words, one of the normal modes became 'soft', that is its frequency became zero, at the transition temperature. This means that the static displacements associated with the existence of ferroelectricity are not simply a sum of basis vectors of a number of irreducible representations of G_1 ; rather *they form a basis of a single irreducible representation* of G_1. At first sight this idea may seem strange, but it should seem much less unreasonable if we recall the relationship that we noted in the previous section between the irreducible representations of a paramagnetic space group G_1 and the Heesch–Shubnikov space groups, G_2, of possible magnetically ordered phases with the atoms in the positions with the symmetry of G_1. Experimental evidence in support of the view that the ferroelectric phase transition is associated with the softening of one of the normal modes of vibration, as a material is cooled through the transition temperature, was first obtained from inelastic neutron scattering work and Raman scattering work on $SrTiO_3$ (Barker and Tinkham 1962, Cowley 1962, Spitzer *et al.* 1962), see fig. 24. Once it was realized that the existence of ferroelectricity could be regarded as resulting from the 'softening' of one of the normal modes of vibration at the transition temperature, then great strides could be made in the understanding of the physics of ferroelectricity ; for further details see, for example, the review article by Cochran (1960).

These are similar to the rules governing the choice of **k** in the representation analysis of magnetic phase transitions in § 8.1. The third possibility which arose in the case of the magnetic ordering, namely when the ordering pattern is incommensurate with the crystal structure, does not occur. This is because the physical origin of the ordered dipole moments is much more intimately related to the positions of the atoms in the case of ferroelectric materials than in the case of magnetically ordered materials.

We should remind ourselves that, although the ferroelectric transition involves displacements of the atoms, we do not regard this as a ‘ drastic structural re-arrangement ’. These displacements are quite small at the transition temperature, although they may become larger at lower temperatures. Thus, it is a continuous phase transition, there being continuous changes in the atomic positions but, of course, a discontinuous change in the symmetry from space group $\mathbf{G_1}$ to space group $\mathbf{G_2}$.

If the space groups of both the paraelectric phase and the ferroelectric phase of a given crystal are known, it is possible to identify **k** for the soft mode. It is also possible to determine the normal-mode eigenvectors for the representations $\Gamma_j{}^{\mathbf{k}}$ of $\mathbf{G_1}$. It is then possible to predict which of the normal modes at **k** will be the soft mode by identifying the particular mode at **k** for which the eigenvector is just the set of displacements that occur in the ferro-electric phase. On the other hand, one may know the space group $\mathbf{G_1}$ of the paraelectric phase and the soft mode may have been identified experimentally by observing its frequency to reduce dramatically towards zero in some experimental work. The space group of the ferroelectric phase could then be identified by determining the displacements of the atoms given by the eigenvector of the soft mode.

We illustrate the above discussion of soft modes and ferroelectric phase transitions by considering the example of the ferroelectric phase transition in $BaTiO_3$ or $SrTiO_3$ (Cochran 1960, Cowley 1962, 1964). The structure of these materials in the paraelectric phase is illustrated in fig. 23 ; the space group $\mathbf{G_1}$ of this structure is $Pm3m$ $(O_h{}^1)$. The details of the group-theoretical analysis of this material are given by Cowley (1964). The ferroelectric phase transition does not involve any multiplication of the volume of the unit cell and therefore the soft mode must be one of the modes at $\mathbf{k} = \mathbf{0}$. The normal modes at Γ $(\mathbf{k} = \mathbf{0})$ can be shown to belong to

$$\Gamma_\xi{}^\Gamma = 4\Gamma_{15} \oplus \Gamma_{25} \tag{8.14}$$

where the notation is that of Bouckaert *et al.* (1936) or to

$$\Gamma_\xi{}^\Gamma = 4\Gamma_4{}^- \oplus \Gamma_5{}^- \tag{8.15}$$

in terms of the labels of Koster *et al.* (1963) for the point group $m3m$. If the space group of the ferroelectric phase is $P4mm$ we see from the compatibility tables between $m3m$ and $4mm$, given by Koster *et al.* (1963), that $\Gamma_4{}^-$ of $m3m$ is compatible with Γ_1 of $P4mm$ but that $\Gamma_5{}^-$ of $m3m$ is not. Thus the soft mode must be one of the $\Gamma_4{}^-$ modes. Unfortunately, four of the five (three-fold degenerate) modes at Γ in $SrTiO_3$ or $BaTiO_3$ are $\Gamma_4{}^-$ modes so that the usefulness of this result in this case is rather limited. The other aspect which we have overlooked is that in $SrTiO_3$ there is not just one phase transition. The ferroelectric phase transition in $SrTiO_3$ is in the vicinity of

30 K (see fig. 24), but there is a cubic to tetragonal phase transition at 110 K ; this is a displacive phase transition and the tetragonal phase has a value of c/a very close to unity. Strictly speaking therefore the phonon symmetries for the material below 110 K but above the ferroelectric phase transition temperature should be described in terms of the representations of the space group of this tetragonal, but non-ferroelectric, phase.

We have previously noted that the onset of magnetic ordering is usually accompanied by a magnetostrictive distortion. This distortion occurs in a manner that is compatible with the reduction of the symmetry associated with the magnetic ordering. There may be several possible choices of direction for the preferred orientation associated with the magnetic ordering, with the resultant occurrence of magnetic domains even within a crystal that was a single crystal in its non-magnetic phase. This formation of domains could be prevented by the application of an external magnetic field to make one of the possible magnetization directions more favourable than the others. A similar thing may happen for the distortions associated with ferroelectric phase transitions, when we have the possibility of ferroelectric domains in a specimen that was a single crystal in the paraelectric phase. The formation of these domains could be prevented by using a large external electric field to make one of the polarization directions more favourable than the others. The concept of ' ferroelasticity ', which was introduced by Aizu (1969), is relevant to the discussion of displacive phase transitions. By the term ' displacive phase transition ' we mean precisely one of those transitions to which we determined to restrict ourselves at the beginning of § 8. Such a phase transition does not involve any drastic structural re-arrangement of the atomic positions, but involves only rather small displacements of the equilibrium positions of the atoms ; such displacements are generally accompanied by a reduction in the symmetry of the crystal. The term ferroelasticity was originally introduced in connection with ferroelectric phase transitions, but it can also be used in connection with magnetic phase transitions (Cracknell 1972). The relevance of the concept of ferroelasticity to magnetic phase transitions can be seen from the fact that the application of mechanical stress to a single-crystal specimen, while it undergoes the transition from the non-magnetic state to the magnetically ordered state, is one of the ways of attempting to produce single-domain single-crystal specimens of magnetic materials. If a magnetic phase transition only involved an ordering of the magnetic moments of the atoms in a crystal, without any associated magnetostrictive distortion, the concept of ferroelasticity would not be relevant to magnetic phase transitions.

There is an analogy between the symmetry studies of the property of ferroelasticity and of the properties of ferroelectricity and of ferromagnetism. We have already seen that it is common to regard a ferroelectric crystal as a crystal in which there exists a spontaneous electric polarization $\mathbf{P}$ in the absence of an external electric field, while in a ferromagnetic crystal there exists a spontaneous magnetization $\mathbf{M}$ in the absence of an external magnetic field. Instead of regarding a ferroelectric or ferromagnetic crystal as a crystal in which there is a non-zero vector, $\mathbf{P}$ or $\mathbf{M}$, there is an alternative approach which is more illuminating in connection with the introduction of the concept of ferroelasticity. If a certain specimen is ferroelectric it can be regarded as being capable of existing in either of two ' orientation states ' with polarizations **P** and −**P** in the absence of an external electric field. It is then possible to ' flip ' the polarization of the specimen from **P** to −**P** (or from −**P** to **P**) by the application and subsequent removal of an appropriate external electric field. Similarly, a ferromagnetic crystal can be regarded as a crystal which is capable of existing in either of two states with magnetization **M** and −**M**, where the magnetization of the specimen can be flipped from one value to the other by a suitable external magnetic field. In the case of ferroelasticity the polarization, **P**, or the magnetization, **M**, is replaced by the mechanical strain which corresponds to the atomic displacements and which is a tensor of rank two. Similarly, the external electric field, or magnetic field, which can be used to ' flip ' a specimen between the two orientation states ±**P**, or ±**M**, is replaced by an external mechanical stress, which is also a tensor of rank 2. In the definition of ferroelasticity (Aizu 1969) : ' A crystal is said to be *ferroelastic* when it has two or more orientation states in the absence of mechanical stress, or of other external fields, and can be made to change from one to another of these states by a mechanical stress '. In addition to magnetic transitions and ferroelectric transitions, ferroelasticity also covers displacive phase transitions which do not involve the appearance of either ferroelectricity or of magnetic ordering. Suppose that a certain point group G_2 describes the symmetry of a specimen of a crystal which can be regarded as obtained by only a small distortion of a structure that possesses the symmetry of another point group G_1, which is a supergroup of G_2 and which is described as the ' prototypic ' point group of G_1 in this structure. G_1 and G_2 may be classical groups or they may be Heesch–Shubnikov groups. Suppose also that one orientation state of a given ferroelastic crystal is labelled as S_1 ; then each of the operations of the supergroup G_1 either regenerates S_1 or generates one of a number of other orientation states S_2, S_3, ... Ferroelasticity will be forbidden if the form of the tensor representing elastic strain is identical for the various orientation states S_1, S_2, S_3, ... An alternative but equivalent approach would be to say that ferroelasticity is forbidden unless the tensor representing elastic strain takes a simpler form in G_1 than in G_2. By ' taking a simpler form ' we mean that there are some components which are required by symmetry to vanish or to be equal to other components (see § 3.1).

The consideration of the property of ferroelasticity from the viewpoint of symmetry studies differs from ferroelectricity and ferromagnetism in one rather important respect. The question of the possibility of the existence of ferroelectricity or of ferromagnetism in a crystal with the symmetry of some given point group is determined by the transformation properties of certain tensors of rank one under the operations of that point group (see § 3.1). The possibility of the existence of ferroelectricity or of ferromagnetism in any given point group **G** is therefore a property of that point group itself and is independent of the point group which describes the symmetry of the para-electric or paramagnetic phase of the crystal and which, generally, is some supergroup of **G**. Therefore in connection with tables such as tables 7 and 8 which list possible symmetries of ferroelectric and ferromagnetic crystals, there is no consideration given to the symmetry of the crystal in the high-temperature phase. However, in the case of ferroelasticity the symmetry restrictions on the existence of ferroelasticity do not simply consist of determining the form of a certain tensor in a given point group. Neither is the property of ferro-elasticity a property of a point group **G** on its own, but rather it is a property of this point group considered in relation to some given prototypic group which is a supergroup of **G**. For example, suppose that a certain cubic material undergoes a displacive phase transition to a tetragonal structure with c/a very close to unity, as occurs, for example, in $BaTiO_3$ or $SrTiO_3$ (see fig. 23). The displacements which occur in the ferroelectric phase are parallel to one of the sets of edges of the cubic unit cell. The common direction of this set of edges survives as a four-fold axis of the crystal, but the displace-ments destroy the other four-fold axes of symmetry and thereby reduce the symmetry of the crystal from cubic ($Pm3m$) to tetragonal ($P4mm$). If one starts with a single-crystal specimen of the cubic phase there will be three possible choices of direction for the four-fold axis of the tetragonal phase and, therefore, also the possibility of the existence of domains. However, suppose that the whole crystal has been induced to produce a single crystal of the tetragonal phase with the four-fold axis along, say, the x axis of the cubic phase. Then if it is possible, by the application of an external mech-anical stress, to 'flip' the specimen's distortion so that the four-fold axis is now along either the y axis or the z axis of the cubic phase the material would provide an example of ferroelasticity. This illustrates the statement that ferroelasticity is not just a property of the point group of the tetragonal phase, but of the fact that this point group arises in this particular example in a material which is only very slightly distorted from a structure with higher symmetry. Thus in constructing a table of point groups in which ferro-elasticity may exist it is, therefore, necessary to specify the appropriate supergroup as well. The details of the identification of the point groups of all possible ferroelastic species has been considered by Aizu (1969, 1973) for the classical point groups (type I Heesch–Shubnikov point groups) and the ideas involved have also been extended to the black and white groups by Aizu (1970) and Cracknell (1972).

8.3. *Jahn–Teller effects*

The 'effect' or group of 'effects' which are now associated with the names of Jahn and Teller have a curious history, at least as far as the physics of solids is concerned. The original papers (Jahn and Teller 1937, Jahn 1938) were concerned exclusively with molecules. Recently, however, it has come to be appreciated that the Jahn–Teller effect for molecules is but one manifestation of a feature which is of substantial fundamental importance and which produces similar effects in other systems as well. In this section we shall be concerned with some of the group-theoretical aspects of the Jahn–Teller effect, in both its original form and in its more general extensions. The more extended use of the term 'Jahn–Teller effect' has only been introduced quite recently and so there are rather few comprehensive reviews or monographs available on the subject ; we would, however, recommend the interested reader to consult the book by Englman (1972). The term 'vibronic coupling' is often used to describe the coupling that we are con-cerned with in this section ; but this term is sometimes restricted to molecular systems or to systems of small groups of atoms with localized states in a solid. To begin with we shall describe some of the features of the original presentation by Jahn and Teller, distinguishing between the general group-theoretical considerations and the particular considerations that apply only to molecules.

The essential group-theoretical ideas involved in the discussion of the Jahn–Teller effect are quite simple and provide another example of the application of Kronecker products in the determination of selection rules. Although the discussion will be formulated in terms of the behaviour of molecules, it can easily be re-formulated for other systems such as small groups of atoms in solids or for large collections of atoms in the form of a large specimen of a crystalline solid. An arbitrary displacement of a molecule can be expanded in terms of the eigenvectors for the normal modes of vibration of the molecule. (We have encountered the related expansion for a solid in eqn. (6.33) previously.) If ξ is a vector representing the displacements of all the atoms the expansion would take the form

$$\xi = \sum_i Q_i \eta_i \qquad (8.16)$$

where η_i are the normal mode eigenvectors, distinguished from one another by the label i. ξ and η_i are vectors in $3n$ dimensions, where n is the number of atoms in the molecule. The Hamiltonian for the motion of an electron, if the ions were frozen in the displaced positions specified by ξ, could then be written in the form

$$\mathscr{H} = \mathscr{H}_0 + V(\mathbf{r}, \xi) \qquad (8.17)$$

where $V(\mathbf{r}, \xi)$ is some function of ξ and of the position, $\mathbf{r}$, of the electron. In practice ξ is not a constant fixed pattern of displacements of the ions, but is a complicated function of time (see eqn. (6.33)). This makes the determination of the eigenvalues and eigenfunctions of $\mathscr{H}$ in eqn. (8.17) a very difficult problem, which is usually simplified by adopting the Born–Oppenheimer approximation. However, in the Jahn–Teller effect we are not concerned with the determination of the eigenvalues and eigenfunctions of $\mathscr{H}$, but only with the rather simple problem of investigating whether a system is stable against a static displacement corresponding to a distortion of the structure. Jahn and Teller then expressed $V(\mathbf{r}, \xi)$ as a power series expansion in η_i, so that

$$\mathscr{H} = \mathscr{H}_0 + \sum_i V_i(\mathbf{r})\eta_i + \sum_{i,j} V_{ij}(\mathbf{r})\eta_i\eta_j + \cdots \qquad (8.18)$$

Since the η_i belongs to some irreducible representation Γ_i of **G**, the point group of the molecule, the functions $V_i(\mathbf{r})$, $V_{ij}(\mathbf{r})$, ... will also belong to various irreducible representations of **G** so that each term on the right-hand side of eqn. (8.18) belongs to the identity representation of **G**.

Suppose that E_0 is an eigenvalue of $\mathscr{H}_0$, in other words E_0 is the energy of one of the electronic states when the ions are in their equilibrium positions, that is when $\xi = \mathbf{0}$. The energy, $E_0 + E'$, of this electronic state for non-zero displacements can be obtained by determining the eigenvalues of $\mathscr{H}$. If E' is calculated by using perturbation theory it will contain terms such as

$$V_{\rho\rho} = \int \phi_\rho^*(\mathbf{r}) V_i(\mathbf{r})\phi_\rho(\mathbf{r}) \, d\mathbf{r} \qquad (8.19)$$

if E_0 is non-degenerate, or

$$V_{\rho\sigma} = \int \phi_\rho{}^*(\mathbf{r}) V_i(\mathbf{r}) \phi_\sigma(\mathbf{r}) \, d\mathbf{r} \qquad (8.20)$$

if E_0 is degenerate, where $\phi_\rho(\mathbf{r})$ and $\phi_\sigma(\mathbf{r})$ are the wave functions corresponding to E_0. The integrals in eqns. (8.19) and (8.20) are of the same form as the matrix elements we have encountered several times already in connection with the determination of selection rules in § 7. If the identity representation Γ_1 (or $\Gamma_1{}^+$) of **G** is not obtained in the reduction of the triple Kronecker product of the representations to which $\phi_\rho{}^*(\mathbf{r})$, $V_i(\mathbf{r})$ and $\phi_\rho(\mathbf{r})$ or $\phi_\sigma(\mathbf{r})$ belong, this integral must vanish. If this integral vanishes for all the normal modes of vibration, the matrix element $V_{\rho\rho}$ or $V_{\rho\sigma}$ will vanish. Therefore, E' vanishes and this means that the system will be stable against an arbitrary static distortion. On the other hand, if Γ_1 (or $\Gamma_1{}^+$) is obtained in the reduction of the triple Kronecker product to which the integral belongs then there is no reason to suppose that E' should be zero; in this case the system is unstable with respect to the arbitrary static distortion ξ.

Suppose that $\phi_\rho(\mathbf{r})$, $\phi_\sigma(\mathbf{r})$ and $V_i(\mathbf{r})$ belong to the irreducible representations Γ_ρ, Γ_σ and Γ_V of **G**. If E_0 is non-degenerate the triple Kronecker product which describes the symmetry of $V_{\rho\rho}$ simplifies to

$$\Gamma_\rho{}^* \boxtimes \Gamma_V \boxtimes \Gamma_\rho = (\Gamma_\rho{}^* \boxtimes \Gamma_\rho) \boxtimes \Gamma_V = \Gamma_V. \qquad (8.21)$$

Therefore, unless $V_i(\mathbf{r})$ belongs to Γ_1 (or $\Gamma_1{}^+$) of **G**, $V_{\rho\rho}$ will vanish and the atomic configuration is stable. If V_i belongs to Γ_1 (or $\Gamma_1{}^+$), that is when the displacement belongs to the totally symmetrical (or breathing) mode, $V_{\rho\rho}$ will not necessarily vanish. However, in this mode the pattern of the displacements of the atoms possesses the full symmetry of **G**. Therefore, as far as non-degenerate energy levels are concerned, the interactions between the electronic states and the atomic vibrations will not cause any distortion that would lower the symmetry of the molecule. If E_0 is degenerate we would expect that

$$\int \phi_\rho{}^*(\mathbf{r}) V_i(\mathbf{r}) \phi_\sigma(\mathbf{r}) \, d\mathbf{r} = \int \phi_\sigma{}^*(\mathbf{r}) V_i(\mathbf{r}) \phi_\rho(\mathbf{r}) \, d\mathbf{r} \qquad (8.22)$$

so that the triple Kronecker product involves a symmetrized square and becomes $[\Gamma_\rho]^2 \boxtimes \Gamma_V$. It then remains to be determined whether the reduction of this product contains Γ_1 (or $\Gamma_1{}^+$) of **G**. The argument, so far, is actually quite general; although it has been restricted, at least by implication, to molecules it could easily be applied to localized states for a group of a small number of atoms in a solid or to extended states in a solid.

Jahn and Teller (1937) took the general result that we have just described, namely that for the stability of a system against an arbitrary distortion the product $[\Gamma_\rho]^2 \boxtimes \Gamma_V$ should not contain Γ_1 (or $\Gamma_1{}^+$), and examined the consequences of this in detail for molecules. The following theorem was enunciated (Jahn and Teller 1937, Jahn 1938):

"A configuration of atoms in a polyatomic molecule will be unstable if the total electronic state (orbital plus spin) is degenerate, unless (i) it is a linear molecule, or (ii) the electronic degeneracy is the Kramers' degeneracy which occurs if the molecule has an odd number of electrons."

This theorem was established by enumeration of all the various possibilities in each of the crystallographic point groups and in various important non-crystallographic point groups.

It would not be unfair to say that for a very long time the full significance of the Jahn–Teller ' effect ' was not appreciated by physicists. Several of the writers of books on the applications of group theory in physics have ignored the effect altogether (Lyubarskii 1960, Hamermesh 1962, Tinkham 1964), while others have given it very scant attention (for example, Heine (1960) relegates the topic to a problem at the end of one of the sections of his book). Other writers, including the present author, have only discussed the Jahn–Teller effect within the limited context of the original statement of the theorem in terms of molecules ; the comment of Meijer and Bauer (1962) that " in 1937 Jahn and Teller discovered the ... interesting and useful theorem ... " is typical (see also Knox and Gold (1964), Cracknell (1968 a)). The failure of physicists to appreciate any deep significance in the theorem of Jahn and Teller was probably partly due to the fact that it had been proved by enumeration, which might have given the impression that it was just an interesting curiosity. Moreover, the fact that the theorem was stated for molecules gave the impression that it was something that was principally of interest to chemists. Now it is realized that ' effects ' similar to that noted by Jahn and Teller also occur in solids, not only for localized states but also, to some extent, involving extended states as well.

In several of the earlier sections we have made use of the Born–Oppenheimer, or adiabatic, approximation in which it is assumed that we can consider separately the vibrational and electronic behaviour of a system. This separation is very useful because many of the physical properties of a solid can be discussed quite satisfactorily either exclusively in terms of the vibrational states or exclusively in terms of the electronic states. Even a quantity such as the specific heat of a metal, which involves both the electrons and the lattice vibrations, can be regarded to quite a good first approximation as a sum of two independent contributions, one from the lattice vibrations and one from the conduction electrons, with no appreciable interaction between them. It is also possible to consider the extension of the Born–Oppenheimer approximation to enable us to achieve a separation between the consideration of spin waves and of lattice vibrations in magnetically ordered crystals. However, there are some physical phenomena in which the interaction between the electrons and the lattice vibrations, that is the electron–phonon inter-action, is important (for example, ultrasonic attenuation, electron spin-lattice relaxation, the electrical resistivity of metals) ; in some cases, notably super-conductivity, the electron–phonon interaction is essential even for the existence of the phenomenon.

Instinctively the first theoretical approach that one adopts to the electron–phonon interaction is to try to treat it by using perturbation theory. However. this is only likely to be a reasonable approach when the interaction energy is small compared with the energies involved when the interactions are neglected. The quanta of vibrational energy are, typically, much smaller than the electronic energies ; therefore what is important is the relative magnitudes of the electron–phonon interaction energy and the energy of a typical phonon in the Born–Oppenheimer approximation. If the electron–phonon interaction, or the correction to the Born–Oppenheimer approximation, is comparable with the vibrational energy, then it becomes necessary to include it in the Hamiltonian of the problem at the outset and to obtain the solution accordingly, instead of just including it afterwards as a perturbation. The original Jahn–Teller effect in molecules was concerned with the possibility of the distortion of a molecule as a result of coupling between electronic and vibrational states. The terms ' Jahn–Teller effect ' and ' pseudo-Jahn–Teller effect ' are now used much more generally to include other situations in which distortions occur in a system as a result of the interaction between the electrons and the vibrations of the atoms. Therefore these terms have come to be applied to systems in which one has to adopt a non-perturbation approach to the coupling. Characteristics of a strong Jahn–Teller effect are

(i) marked changes in the electronic properties,
(ii) great propensity to suffer distortion,
(iii) an inherent stability to maintain the distortion.

It may seem strange to use the term Jahn–Teller effect in a situation when the electron-vibration coupling is so large that one has to adopt a non-perturbation theory approach, whereas the discussion around eqns. (8.19) and (8.20) is all based on the perturbation theory matrix elements. However, this accurately reflects the change in emphasis that is felt to reflect the physically important aspect of the effect. In this situation group-theoretical attention has to be given to the terms in the Hamiltonian $\mathscr{H}$ in eqn. (8.18) rather than to the perturbation matrix elements in eqns. (8.19) and (8.20). However, the symmetry of an eigenstate cannot change discontinuously as the magnitude of the coupling parameter is varied, provided the group **G** of the symmetry operations of the system is not changed. Therefore, group-theoretical results which were obtained on the basis of perturbation theory in the situation of weak coupling may remain valid when the coupling becomes stronger.

The discussion of the Jahn–Teller effect for localized states in solids is very similar to that for molecules. The symmetries of the localized states will be described in terms of the irreducible representations of one of the point groups and the group-theoretical analysis proceeds just as for molecules. The systems involved may be point defects, interstitial or substitutional impurity atoms, or small groups or clusters of atoms. The experimental observation of the effect is likely to be made indirectly, namely from spectroscopic observations of splittings of electronic levels as a result of the Jahn–Teller distortion. For extended states, however, the approach to the Jahn–Teller effect is necessarily rather different. First, the symmetries of the extended states have to be described in terms of the irreducible representations of a space group and not of a point group. Secondly, any distortion which occurs as a result of (non-perturbational) electron–phonon coupling must be such that its translational symmetry is compatible with one of the space groups. If the distortion is absent at high temperatures but exists below a certain critical temperature, T_c, we have a displacive phase transition. The Jahn–Teller effect has therefore been studied quite extensively as a possible cause of displacive phase transitions, both for displacive phase transitions that involve the appearance of ferroelectricity and for those which do not involve ferroelectricity. Any distortion which occurs in a displacive phase transition is a state of the whole crystal. Its occurrence and the possible displacements which can occur can be discussed in terms of representation analysis in a manner that is similar to that used in § 8.1 for magnetic structures.

Any distortion which occurs in a displacive phase transition must be a state of the whole crystal and whether it is stable will now not be determined just by a small number of matrix elements $V_{\rho\rho}$ or $V_{\rho\sigma}$ but by the sum of a very large number of matrix elements associated with a very large number of electron and phonon states. Group-theoretical arguments analogous to those used for molecules can be developed for the matrix elements involving extended states in solids (Birman 1962 a, Kristofel' 1964).

Suppose that $\psi_m{}^{\mathbf{k}_p}(\mathbf{r})$ is the wave function of an electron with wave vector $\mathbf{k}_p$ and belonging to the irreducible representation $\Gamma_m{}^{\mathbf{k}_p}$ of $\mathbf{G}^{\mathbf{k}_p}$. The Hamiltonian for an arbitrary displacement, $\boldsymbol{\xi}$, which can be resolved in terms of normal-mode eigenvectors (see eqn. (6.33)) can be written in the form of eqn. (8.17) and the potential $V(\mathbf{r}, \boldsymbol{\xi})$ can be expanded in terms of the normal mode eigenvectors $\boldsymbol{\eta}(\mathbf{k}_q, n)$. Thus

$$\mathscr{H} = \mathscr{H}_0 + \sum_{n, p} V_{nq}(\mathbf{r})\boldsymbol{\eta}(\mathbf{k}_q, n)$$
$$+ \sum_{n, p\, ;\, n', p'} V_{nqn'q'}(\mathbf{r})\boldsymbol{\eta}(\mathbf{k}_q, n)\boldsymbol{\eta}(\mathbf{k}_{q'}, n') + \dots \quad (8.23)$$

Since $\boldsymbol{\eta}(\mathbf{k}_q, n)$ belongs to $\Gamma_n{}^{\mathbf{k}_q}$ the potential $V_{nq}(\mathbf{r})$ also belongs to $\Gamma_n{}^{\mathbf{k}_q}$ (or to $\Gamma_n{}^{\mathbf{k}_q*}$ if $\Gamma_n{}^{\mathbf{k}_q}$ is complex). As in the molecular case, the configuration will be stable if the linear terms in eqn. (8.23) vanish, that is if the triple product $[\Gamma_m{}^{\mathbf{k}_p}]^2 \boxtimes \Gamma_n{}^{\mathbf{k}_q}$ does not contain the identity representation Γ_1 (or $\Gamma_1{}^+$), at $\mathbf{k} = \mathbf{0}$, of the space group $\mathbf{G}$. If $\Gamma_m{}^{\mathbf{k}_p}$ is a non-degenerate representation of $\mathbf{G}^{\mathbf{k}_p}$ then $[\Gamma_m{}^{\mathbf{k}_p}]^2$ will necessarily correspond to just Γ_1 (or $\Gamma_1{}^+$) of $\mathbf{G}$ so that the triple product $[\Gamma_m{}^{\mathbf{k}_p}]^2 \boxtimes \Gamma_n{}^{\mathbf{k}_q}$ will only contain Γ_1 (or $\Gamma_1{}^+$) if $\Gamma_n{}^{\mathbf{k}_q}$ is Γ_1 (or $\Gamma_1{}^+$), i.e. only for a set of displacements corresponding to the totally symmetrical mode of vibration belonging to Γ_1 (or $\Gamma_1{}^+$) at $\mathbf{k} = \mathbf{0}$. This mode does not cause any distortion that would lower the symmetry of the crystal. Therefore, just as in the case of molecules, it is only degenerate electronic states that are of relevance in the Jahn–Teller effect. Birman (1962 a) applied these arguments to the points Γ, L and X in the Brillouin zones of the structures of diamond and zinc blende (see fig. 8); his results are reproduced in table 39. Of all the degenerate electronic states at Γ, L and X only $\Gamma_{12}{}^\pm$ in $Fd3m$ and Γ_{12} in $F\bar{4}3m$ are stable against all lattice distortions. As an extrapolation from the results for these two structures Birman (1962 a) suggested a general result for all space groups : '' almost all degenerate electronic states of a crystal are configurationally unstable, with a consequent crystal distortion and concomitant splitting of the degeneracy to be expected.'' The splitting of the degeneracy may, of course, be rather small and consequently difficult to observe ; this appears to be the case in both the diamond and zincblende structures. Nevertheless there are some other structures in which the Jahn–Teller effect has been seriously considered as the cause of displacive phase transitions ; these include perovskite-type structures (see also § 8.2), V_3Si-type structures, and some rare-earth vanadates (see Elliott (1971) and chapter 7 of the book by Englman (1972)).

Table 39.   Normal-mode distortions for which the electronic states are unstable in
diamond and zinc blende structures

Electronic state	Degeneracy	Normal mode distortion	Type
	$Fd3m\ (O_h{}^7)$		
$\Gamma^{(15\pm)}$; $\Gamma^{(25\pm)}$	3	$\Gamma^{(25+)}$	Optic
$\star X^{(1)}$; $\star X^{(2)}$; $\star X^{(3)}$; $\star X^{(4)}$	6	$\Gamma^{(25+)}$ $\star X^{(1)}$ $\star X^{(4)}$	Optic $LA+LO$ TO
$\star L^{(1\pm)}$; $\star L^{(2\pm)}$	4	$\Gamma^{(25+)}$ $\star X^{(1)}$	Optic $LA+LO$
$\star L^{(3\pm)}$	8	$2\Gamma^{(25+)}$ $2\star X^{(1)}$ $\star X^{(4)}$	Optic $LA+LO$ TO
	$F\bar{4}3m\ (T_d{}^2)$		
$\Gamma^{(15)}$; $\Gamma^{(25)}$	3	$\Gamma^{(15)}$	Optic (Reststrahl)
$\star X^{(1)}$; $\star X^{(2)}$	3	$\star X^{(1)}$	LO (or LA)
$\star X^{(3)}$; $\star X^{(4)}$	3	$\star X^{(3)}$	LA (or LO)
$\star X^{(5)}$	6	$\Gamma^{(15)}$ $\star X^{(1)}$ $\star X^{(3)}$ $\star X^{(5)}$	Optic LO (or LA) LA (or LO) TO and TA
$\star L^{(1)}$; $\star L^{(2)}$	4	$\Gamma^{(15)}$ $\star X^{(1)}$ $\star X^{(3)}$	Optic LO (or LA) LA (or LO)
$\star L^{(3)}$	8	$2\Gamma^{(15)}$ $2\star X^{(1)}$ $2\star X^{(3)}$ $\star X^{(5)}$	Optic LO (or LA) LA (or LO) TO and TA

Note.   Only those representations which correspond to normal modes of $Fd3m$
$(O_h{}^7)$ or $F\bar{4}3m$ $(T_d{}^2)$ are listed in column 3.
$\star\mathbf{k}_j$ indicates the star of $\mathbf{k}_j$.   (Birman 1962 a.)

In spite of the discussion of the general theory and of the examples of the
diamond and zinc blende structures in the previous paragraph, it should be
remembered that the special points of symmetry are vanishingly small in
number, compared with the points with general wave vectors. Therefore,
any group-theoretical results associated with the special points of symmetry
will be of little significance, because the contributions from these points to
the total energy of the crystal will be swamped by the contributions from the
points with general wave vectors. Another complication that we have not

considered at all is the possibility of going beyond the linear terms in the expansions in eqn. (8.18) and eqn. (8.23); this of course applies to the cases of both the localized and the extended states. If we do proceed beyond the linear approximation the Kronecker products that have to be reduced will be the products of four, or more, irreducible representations.

8.4. *Landau's theory of second-order phase transitions*

For a phase transition which does not involve a drastic structural re-arrangement it is common to be able to find a subgroup relation between the space groups G_1 and G_2 of the two phases. Such phase transitions may be displacive transitions, magnetic transitions, or ferroelectric transitions. In a displacive transition the low temperature phase, with space group G_2, can be regarded as a slightly distorted form of the high temperature phase, with space group G_1, where the magnitude of the distortion either tends asymptotically to zero at the transition temperature or, at least, is very small at the transition temperature. There is consequently a continuous change in the structure of the crystal but a discontinuous change in the symmetry of the crystal. G_2, however, will be expected to be a subgroup of G_1. In a transition from a paramagnetic phase, with a (grey) space group M_1, to a magnetically ordered phase, with a (magnetic) space group M_2 we would

Fig. 25

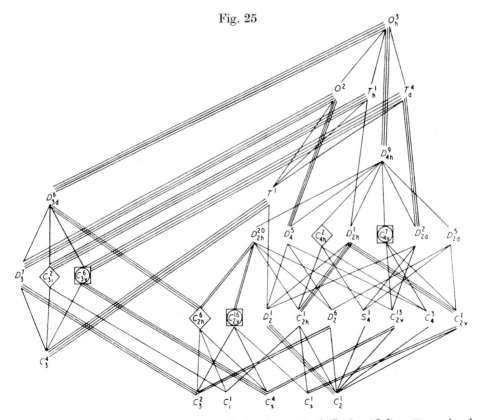

Lattice of subgroups (with same translational part) of $Pm3n$ (O_h^3); □ maximal **P**-subgroups; ◇ maximal **M**-subgroups; ○ maximal **j**-subgroups (Ascher 1967).

also expect $\mathbf{M}_2$ to be a subgroup of $\mathbf{M}_1$. Similarly, in a ferroelectric transition the (black and white) space group $\mathbf{M}_2$ of the ferroelectric phase will be a subgroup of the (grey) space group $\mathbf{M}_1$ of the paraelectric phase.

Very often the space group $\mathbf{G}_1$ of one phase (usually the high-temperature phase) is known but the space group $\mathbf{G}_2$ of the other phase is not known; the subgroup relation enables many space groups to be eliminated as candidates for $\mathbf{G}_2$. For example at 20·5 K, which is about 3·5 K above its superconducting transition temperature, V_3Si undergoes a displacive phase transition from a cubic structure, with $\mathbf{G}_1 = Pm3n$ $(O_h{}^3)$, to a tetragonal structure with c/a very close to 1. All the subgroups of $Pm3n$ $(O_h{}^3)$ which have the same translational subgroup are identified, in the Schönflies notation, in fig. 25. The term 'zellengleich' is sometimes applied to space groups having the same translational subgroup, and diagrams similar to fig. 25 for the other classical space groups have been constructed by Ascher (1968).

In assuming that $\mathbf{G}_1$ and $\mathbf{G}_2$ both have the same translational subgroup we are assuming that there is no change in the size of the unit cell of the crystal; in other words the distortion of the crystal is compatible with the translational symmetry of the space group $\mathbf{G}_1$ and does not cause any of the basic translation vectors to be multiplied by an integer. The fact that for the example of this transition in V_3Si the space group $\mathbf{G}_2$ is known to be tetragonal, means that of all the subgroups of $Pm3n$ $(O_h{}^3)$ in fig. 25 the only ones that are candidates as $\mathbf{G}_2$ for V_3Si are

$$
\left.
\begin{array}{ll}
P4_2/mmc\ (D_{4h}{}^9) & P\bar{4}m2\ (D_{2d}{}^5) \\[2ex]
P4_222\ (D_4{}^5) & P4_2/m\ (C_{4h}{}^2) \\[2ex]
P\bar{4}2c\ (D_{2d}{}^2) & P\bar{4}\ (S_4{}^1) \\[2ex]
P4_2mc\ (C_{4v}{}^7) & P4_2\ (C_4{}^3)
\end{array}
\right\}.
\qquad (8.24)
$$

Bearing in mind the fact that there are altogether 68 tetragonal space groups the use of the subgroup relation clearly achieves a very substantial reduction in the number of possibilities that have to be considered.

For dealing with magnetic phase transitions and ferroelectric transitions, one would need to use the subgroup relations for the Heesch–Shubnikov groups. Tables of the subgroup relations for the Heesch–Shubnikov point groups have been published by Ascher and Janner (1965) and by Cracknell (1974 a). Tables of subgroup relations for the Heesch–Shubnikov space groups are probably not worth compiling, because they would be very bulky. Moreover, most of the information about the subgroup relation between two Heesch–Shubnikov space groups $\mathbf{M}_1$ and $\mathbf{M}_2$ can be obtained by using the subgroup tables for the classical, or Fedorov, space groups which are the unitary subgroups, $\mathbf{G}_1$ and $\mathbf{G}_2$, of $\mathbf{M}_1$ and $\mathbf{M}_2$, respectively.

For a transition between two different magnetically-ordered phases there may not be a subgroup relation between the space groups $\mathbf{M}_1$ and $\mathbf{M}_2$ of the two phases. However, there is likely to be a large number of symmetry operations that are common to both space groups; in other words the intersection $\mathbf{M}_1 \cap \mathbf{M}_2$ is not a trivial space group. Alternatively, each of the space groups $\mathbf{M}_1$ and $\mathbf{M}_2$ may be regarded as a subgroup of the space group $\mathbf{M}_0$

that describes the positions of the atoms, ignoring their magnetic moments. This relationship between M_1 and M_2 via a common supergroup M_0 may also enable some space groups to be eliminated as candidates for M_2.

In addition to the information which can be obtained from the subgroup condition, it may also be possible to obtain some additional information about the symmetry aspects of a phase transition by using arguments based on thermodynamics. These arguments originate from the work of Landau (1937 a, b), which has subsequently been developed by a number of other workers. It would not be appropriate to give a detailed discussion of the thermodynamical arguments here (for details see, for example, chapter 14 of Landau and Lifshitz (1958)); we shall simply note the group-theoretical aspects of the results.

It is conventional in thermodynamics to classify phase transitions in a manner that was introduced by Ehrenfest (see, for example, chapter 9 of the book by Pippard (1957)). In this classification the order of the lowest derivatives of the Gibbs function which are discontinuous at the phase transition is taken to define the order of the phase transition. The classification scheme for second-order and higher-order phase transitions is quite elaborate and often quite difficult to apply in practice. It is common now simply to classify any given phase transition as either a first-order transition or a continuous transition. In a first-order transition the first derivatives of the Gibbs function have discontinuities at the transition, so that such a transition is accompanied by a change in entropy, that is by a latent heat, and also by a change in volume. For a continuous transition the first derivatives remain continuous, so that there is no latent heat or change in volume, and only higher-order derivatives, such as the compressibility or the specific heat, are discontinuous. A transition which involves no major structural re-arrangement, and for which therefore a group–subgroup relation exists between the space groups of the two phases, may be a discontinuous or continuous (i.e. first-order or second-order) phase transition, whereas a transition which does involve a drastic structural re-arrangement and for which no subgroup relation exists must necessarily be a discontinuous (i.e. first-order) transition.

The thermodynamical arguments given by Landau apply to continuous phase transitions. These arguments can be manipulated into a set of group-theoretical conditions between the space groups G_1 and G_2 of the material above the transition temperature and below the transition temperature, respectively. For a mathematical description of continuous phase transitions it is convenient to introduce an order parameter, η, which characterizes the extent to which the configuration of atoms, or spins, departs from their configuration in the high-symmetry phase. In the high-symmetry phase $\eta = 0$. The necessary, but not sufficient, conditions for a transition between phases described by the classical space groups G_1 and G_2 to be a continuous transition can be summarized as follows, where Γ_i is the irreducible representation of G_1 to which η belongs (see, for example, Lyubarskii (1960)):

   (i) G_2 is a subgroup of G_1.

   (ii) $[\Gamma_i]^3$ must not contain the identity representation of G_1.

   (iii) $\{\Gamma_i\}^2$ must not contain any representation which has any component of a polar vector as its basis.

   (iv) Γ_i must be compatible with the identity representation of G_2.

Condition (i) has already been discussed on general symmetry grounds without any reference to thermodynamics. A fifth condition which is not really part of the Landau theory but is only a reflection of the fact that η belongs to Γ_i, is used implicitly in most physical arguments in this connection (Birman 1966 b).

(v) Γ_i corresponds to a physical tensor field.

To these conditions we may add the *chain subduction criterion* which was introduced by Goldrich and Birman (1968). We suppose that the representation Γ_i of $\mathbf{G}_1$ satisfies conditions (i)–(v) above and that, in addition, Γ_i of $\mathbf{G}_1$ subduces into Γ_1 (or Γ_1^+) of $\mathbf{G}_2$ *once*. If we now consider a group $\mathbf{G}_2'$ which is a subgroup of $\mathbf{G}_2$ and if Γ_1 of $\mathbf{G}_1$ also subduces into Γ_1 (or Γ_1^+) of $\mathbf{G}_2'$ *once*, then the transition $\mathbf{G}_1 \rightarrow \mathbf{G}_2'$ is eliminated. Expressed mathematically the *chain subduction criterion* is (Goldrich and Birman 1968):

(vi) If

$$\mathbf{G}_1 \supset \mathbf{G}_2 \supset \mathbf{G}_2'$$

and

$$\Gamma_i \downarrow \mathbf{G}_2 \text{ contains } \Gamma_1 \text{ (or } \Gamma_1^+) \text{ of } \mathbf{G}_2 \text{ } once$$

and

$$\Gamma_i \downarrow \mathbf{G}_2' \text{ contains } \Gamma_1 \text{ (or } \Gamma_1^+) \text{ of } \mathbf{G}_2' \text{ } once,$$

then the transition $\mathbf{G}_1 \rightarrow \mathbf{G}_2'$ cannot be a continuous transition. The use of this subduction criterion was illustrated by Goldrich and Birman (1968) to determine the possible space groups of a crystal which starts with the perovskite structure and undergoes a continuous phase transition.

Two theorems involving p, the index of the subgroup $\mathbf{G}_2$ in $\mathbf{G}_1$, are given by Landau and Lifshitz (1958):

(I) " For every transition involving the halving of the number of symmetry operations of a crystal, that is $p = |\mathbf{G}_1|/|\mathbf{G}_2| = 2$, continuous phase transitions can exist."

This theorem is proved in § 136 of Landau and Lifshitz (1958).

(II) " No second-order phase transition can exist for transitions involving the decrease by a factor three of the number of symmetry operations of a crystal, that is, $p = 3$."

This theorem is only stated without any proof by Landau and Lifshitz with the comment that " it appears that the ... theorem is ... true ". A direct proof of theorem II for the particular case of a cubic to tetragonal phase change, such as the one which occurs at 20·5 K in V_3Si, was given by Anderson and Blount (1965). It seems unlikely that there is a general proof of theorem (II) or that this theorem can be extended to $p = 4$, 5, 6, etc.

Applications of conditions (i)–(vi) to several examples of non-magnetic phase transitions will be found in the literature. For example, Lyubarskii (1960) considered the derivation of all the space groups which may arise in a crystal which starts with the space group $I4_1/a$ as $\mathbf{G}_1$ and undergoes a continuous phase transition. Eleven space groups are possible and the details of the derivation are rather complicated; the reader who is interested is referred to § 38 of Lyubarskii (1960). These conditions have also been applied to the prediction of possible space groups $\mathbf{G}_2$ for the structure of V_3Si

at temperatures below 20·5 K at which it undergoes a displacive phase transition, which is known to be a second-order phase transition, and above which G_1 is known (Birman 1966 b, 1967, Perel *et al.* 1968). We have already noted that for this transition the subgroup condition restricts G_2 to being one of eight possible space groups, see (8.24). Since the unit cell is unchanged as a result of the transition, the representation Γ_i of G_1 must be at $k = 0$, i.e. effectively Γ_i is a representation of the point group $m3m$. By using condition (iv) and the appropriate compatibility tables (for example, Koster *et al.* (1963)) it is possible to reject many of the representations of $m3m$ (O_h), see table 40. Using condition (ii) it is possible to show that Γ_1^+, Γ_3^+ and Γ_5^+ are unacceptable as Γ_i. This eliminates the possibility of $P4_2/mmc$. So far we have not made any use of the fact that Γ_i must be the

Table 40. Representations of G_1 compatible with Γ_1 or Γ_1^+ of G_2

	Γ of G_1 (at $k=0$)	Γ_1 or Γ_1^+ of G_2 (at $k=0$)	
1	Γ_1^+, Γ_3^+	Γ_1^+	$P4_2/mmc$
2	Γ_1^+, Γ_3^+, Γ_1^-, Γ_3^-	Γ_1	$P4_222$
3	Γ_1^+, Γ_3^+, Γ_4^-	Γ_1	$P4_2mc$
4	Γ_1^+, Γ_3^+, Γ_2^-, Γ_3^-	Γ_1	$P\bar{4}2c$
5	Γ_1^+, Γ_3^+, Γ_5^-	Γ_1	$P\bar{4}m2$
6	Γ_1^+, Γ_3^+, Γ_4^+	Γ_1^+	$P4_2/m$
7	$\left.\begin{array}{l}\Gamma_1^+, \Gamma_3^+, \Gamma_4^+\\ \Gamma_1^-, \Gamma_3^-, \Gamma_4^-\end{array}\right\}$	Γ_1	$P4_2$
8	$\left.\begin{array}{l}\Gamma_1^+, \Gamma_3^+, \Gamma_4^+\\ \Gamma_1^-, \Gamma_3^-, \Gamma_4^-\end{array}\right\}$	Γ_1	$P\bar{4}$

Note. The representations of G_1 are labelled in the notation of Koster *et al.* (1963) for the point group $m3m$ (O_h). (Adapted from Birman (1966 b, 1967).)

representation to which the order parameter, η, belongs. Since we are concerned with a displacive transition, the transition can be regarded as the sudden 'freezing' of the crystal in one of its $k = 0$ normal modes of vibration, see § 8.2. The acoustic modes may be excluded because they must survive in the lower symmetry phase as well so that the order parameter, η, must belong to one of the irreducible representations to which the $k = 0$ optic modes of the lattice vibrations belong (for further discussion of this see, for example, Sirotin and Mikhel'son (1968)). The problem of the identification of the symmetries of the normal modes of vibration of a crystal has been discussed in § 6.3 ; we simply quote the result for V_3Si, namely that the $k = 0$ optic modes belong to $\Gamma_2^+ \oplus \Gamma_3^+ \oplus \Gamma_4^+ \oplus \Gamma_5^+ \oplus 2\Gamma_4^- \oplus 2\Gamma_5^-$, where the labels are those of Koster *et al.* (1963) for the point group $m3m$ (O_h). Of these normal modes $\Gamma_3{}^+$ and $\Gamma_5{}^+$ are already disallowed by condition (ii) which leaves only $\Gamma_2{}^+$, $\Gamma_4{}^+$, $\Gamma_4{}^-$ and $\Gamma_5{}^-$. This enables us to eliminate $P4_222$ and $P\bar{4}2c$ leaving only space groups 3, 5, 6, 7 and 8 of table 40. By using the chain subduction criterion, Birman (1967) was able to show that $P4_2$ and $P\bar{4}$ can also be eliminated. This leaves the space groups $P4_2mc$, $P\bar{4}m2$ and $P4_2/m$ and actual suggested positions for the V and Si atoms in the structure are given in table 41.

Table 41.   Predicted tetragonal space groups for V_3Si

Site	Symmetry	Coordinates	Atom
		$\mathbf{G}_2 :\ P4_2mc\ (C_{4v}{}^7)$	
c	$mm2\ (C_{2v})$	$(0, \frac{1}{2}, \frac{1}{2}+u)$, $(\frac{1}{2}, 0, u)$	Si
a	$mm2\ (C_{2v})$	$(0, 0, \frac{1}{4}+v)$, $(0, 0, \frac{3}{4}+v)$	V
e	$m\ (C_s)$	$(\frac{1}{4}, \frac{1}{2}, w)$, $(\frac{1}{2}, \frac{1}{4}, \frac{1}{2}+w)$ $(\frac{3}{4}, \frac{1}{2}, w)$, $(\frac{1}{2}, \frac{3}{4}, \frac{1}{2}+w)$	V
		$\mathbf{G}_2 :\ P\bar{4}m2\ (D_{2d}{}^5)$	
g	$mm2\ (C_{2v})$	$(0, \frac{1}{2}, \frac{1}{4}+p)$, $(\frac{1}{2}, 0, \frac{3}{4}-p)$	Si
a	$\bar{4}2m\ (D_{2d})$	$(0, 0, 0)$	V
d	$\bar{4}2m\ (D_{2d})$	$(0, 0, \frac{1}{2})$	V
k	$m\ (C_s)$	$(\frac{1}{4}, \frac{1}{2}, \frac{3}{4}+q)$, $(\frac{1}{2}, \frac{1}{4}, \frac{1}{4}-q)$ $(\frac{3}{4}, \frac{1}{2}, \frac{3}{4}+q)$, $(\frac{1}{2}, \frac{3}{4}, \frac{1}{4}-q)$	V
		$\mathbf{G}_2 :\ P4_2/m\ (C_{4h}{}^2)$	
d	$2/m\ (C_{2h})$	$(0, \frac{1}{2}, \frac{1}{2})$, $(\frac{1}{2}, 0, 0)$	Si
e	$\bar{4}\ (S_4)$	$(0, 0, \frac{1}{4})$, $(0, 0, \frac{3}{4})$	V
j	$m\ (C_s)$	$(\frac{1}{4}+u, \frac{1}{2}+v, 0)$, $(-\frac{1}{2}-v, \frac{1}{4}+u, \frac{1}{2})$ $(-\frac{1}{4}-u, -\frac{1}{2}-v, 0)$, $(\frac{1}{2}+v, -\frac{1}{4}-u, \frac{1}{2})$	V

(Adapted from Birman (1966 b, 1967).)

An alternative approach to the consideration of continuous phase transitions has been suggested by Ascher (1966, 1967) and illustrated for various examples. Although we know that in a continuous transition the group $\mathbf{G}_2$ is a subgroup of $\mathbf{G}_1$ it does not follow that all the subgroups of $\mathbf{G}_1$ can be achieved by means of a continuous transition. The method suggested by Ascher for ferroelectric phase transitions is based on the following proposition :

"The space group, $\mathbf{G}_2$, of a phase that arises in a ferroelectric phase transition is a maximal polar subgroup of the space group, $\mathbf{G}_1$, of the high temperature phase."

By a polar subgroup is meant a space group that allows the existence of a non-zero spontaneous electric dipole moment in the crystal (see § 3.1). A maximal polar subgroup is therefore a polar subgroup that is not contained in any other polar subgroup. The reason given by Ascher for the requirement of a maximal subgroup is on energetic grounds in that it ensures the formation of the minimum number of domain walls ; the use of maximal subgroups is clearly related to, but not identical with, the chain subduction criterion. However, it appears that no general proof of the validity of Ascher's proposition has been published. Examples of the application of this approach have been considered for several ferroelectric transitions by Ascher (1966). The proposition can also be extended in an obvious manner to apply to magnetic transitions.

The group-theoretical conditions (i)–(vi) given above were expressed in terms of the representations of classical space groups, or type I Heesch–Shubnikov space groups. It is possible to apply Landau's theory of continuous phase transitions to magnetic phase transitions. For a transition from a paramagnetic phase to a magnetically ordered phase, $\mathbf{G}_1$ will be a grey space group which includes θ, the operation of time-reversal, as an element of the group. On the other hand, $\mathbf{G}_2$, the space group of the magnetically ordered phase will be a type I, type III, or type IV Heesch–Shubnikov space group and will not contain θ on its own as an element of the group ; compound operations consisting of a product of θ with a space-group operation may be present in $\mathbf{G}_2$.

If we consider any phase transition of a crystal between a magnetically ordered phase and a paramagnetic phase with the same crystallographic structure, there is a general argument due to Landau (see, for example, p. 445 of Landau and Lifshitz (1958)) which shows that such a transition always has the possibility of being a continuous phase transition. For a magnetic transition we can use $\mathbf{M}(\mathbf{r})$, the magnetization, as the order parameter. Because the sign of $\mathbf{M}(\mathbf{r})$ is changed by θ whereas $G(P, T, \eta)$, the Gibbs function, is unaffected by the operation of time-reversal, it was argued by Landau that in the expansion of the Gibbs function in powers of $\mathbf{M}(\mathbf{r})$ all terms with odd powers of $\mathbf{M}(\mathbf{r})$ should be identically zero and therefore all transitions between a paramagnetic phase and a magnetically-ordered phase with the same crystallographic structure have the possibility of being continuous transitions. It should be emphasized that although the expansion of $G(P, T, \eta)$ is invariant under θ not only above T_c but also below T_c, θ is not an element of the group $\mathbf{G}_2$ of the magnetically-ordered phase ; the invariance of $G(P, T, \eta)$ under θ arises because for both phases we have used an expansion with the symmetry of $\mathbf{G}_1$ although such an expansion may be incomplete below T_c (Cracknell et al. 1970, Cracknell 1971 b). If one makes certain simplifying physical assumptions about the forms of the terms that are actually present in $G(P, T, \eta)$ for the magnetically ordered phase then this invariance of $G(P, T, \eta)$ under the operations of $(\mathbf{G}_1 - \mathbf{G}_2)$, and in particular under θ itself, may indeed be preserved below T_c ; for example, terms of the form $\mathbf{M} . \mathbf{B}_i$ are invariant under time-reversal where $\mathbf{M}$ is the (sublattice) magnetization and $\mathbf{B}_i$ is an internal magnetic field within the crystal. However, there are no general grounds for assuming that any given observable may possess more symmetry than that of the crystal itself, although such extra symmetry can always appear by accident. The application of the Landau theory of continuous phase transitions to transitions between the paramagnetic phase and a spiral magnetically ordered phase with the same crystallographic structure has been considered by a number of authors (Dzialoshinskiĭ 1964, Kovalev 1965 b, Sólyom 1966, 1971).

In addition to transitions between a paramagnetic phase and a magnetically ordered phase, there are also magnetic phase transitions, such as spin-flip transitions and metamagnetic transitions, between two different magnetically ordered phases with the same crystallographic structure. For such a transition $\mathbf{G}_1$ will not be a grey space group (type II Heesch–Shubnikov group) and will therefore not contain θ, the operation of time-inversion, on its own, although θ may of course still be present in combination with some space-group operation $\{R|\mathbf{v}\}$. Therefore, for a transition between two different magnetically ordered phases with the same crystal structure there is no general argument that the transition may be a continuous one ; each particular transition of this type has to be investigated individually, either experimentally or theoretically, if one wishes to determine whether it is a continuous transition or not. The thermodynamical results which were expressed in group-theoretical terms in conditions (i)–(vi) above, and which can be used to predict whether or not a given phase transition can be continuous, can be adapted to the case of transitions between two different magnetically ordered phases (see, for example, Cracknell (1971 b, 1974 a) and Backhouse (1974)). This adaptation involves rewriting conditions (i)–(vi) in terms of the co-representations of the Heesch–Shubnikov groups, $\mathbf{M}_1$ and $\mathbf{M}_2$, of the two magnetically ordered phases. For those transitions in which the unit cell is the same (neglecting magnetostrictive distortions) in each of the two phases there is no serious problem involved in the application of these modified conditions (Cracknell 1971 b, Cracknell and Sedaghat 1972). However, in many transitions the volume of the unit cell in one phase may be an integral multiple of the volume of the unit cell in the other phase, or the magnetic ordering pattern may even be incommensurate with the translational symmetry of the crystal, in which case the whole idea of the definition of a unit cell becomes questionable. In these circumstances it is frequent to find that the subgroup condition is violated. It is then tempting to say that this violation of the subgroup condition automatically means that the transition must be a discontinuous (i.e. first-order) transition. However, this is probably a too naïve view of the situation and it seems likely that it is necessary to examine in detail the relationship between $\mathbf{M}_1$ and $\mathbf{M}_2$ that exists via a common supergroup such as $\mathbf{M}_0$, which is the (grey) space group of the atomic positions neglecting the magnetic moments of the atoms. The details of this problem are still unresolved at present.

§ 9. Conclusion

The bulk of this article has been concerned with describing various different examples of the applications of group theory in solid-state physics. We have tried, as we went along, to indicate the relative importance of these various applications. To try to draw any further conclusions now about specific applications, either at present or in the future, would involve a considerable amount of repetition and would probably throw little fresh light on the situation. In the end each application will be judged to be important or unimportant by each individual reader, depending on whether it appears to be useful or not. The only general conclusion that we should perhaps draw, is to say that in the last decade or two we have seen a great expansion of the use of group theory into fresh applications in solid-state physics and that it would be rash to predict at this stage that no new applications will be found in the future. We look forward with interest to see what these new applications will be.

REFERENCES

ABRAGAM, A., and BLEANEY, B., 1970, *Electron Paramagnetic Resonance of Transition Ions* (Oxford : Clarendon Press).
AIZU, K., 1969, *J. phys. Soc. Japan*, **27**, 387 ; 1970, *Phys. Rev.* B, **2**, 754 ; 1973, *J. phys. Soc. Japan*, **35**, 180.
AKHIEZER, A. I., BAR'YAKHTAR, V. G., and PELETMINSKII, S. V., 1968, *Spin Waves* (Amsterdam : North-Holland).
ALTMANN, S. L., 1957, *Proc. Camb. phil. Soc. math. phys. Sci.*, **53**, 343.
ALTMANN, S. L., and BRADLEY, C. J., 1963, *Phil. Trans. R. Soc.* A, **255**, 199 ; 1965, *Rev. mod. Phys.*, **37**, 33.
ALTMANN, S. L., and CRACKNELL, A. P., 1965, *Rev. mod. Phys.*, **37**, 19.
ANDERSON, P. W., and BLOUNT, E. I., 1965, *Phys. Rev. Lett.*, **14**, 217.
ASCHER, E., 1966, *Phys. Lett.*, **20**, 352 ; 1967, *Chem. Phys. Lett.*, **1**, 69, 246 ; 1968, *Lattices of Equi-translation Subgroups of the Space Groups* (Geneva : Battelle Institute).
ASCHER, E., and JANNER, A., 1965, *Acta crystallogr.*, **18**, 325.
ASHBY, N., and MILLER, S. C., 1965, *Phys. Rev.* A, **139**, 428.
ASTROV, D. N., 1960, *Zh. éksp. teor. Fiz.*, **38**, 984. English translation : *Soviet Phys. JETP*, **11**, 708.
BACKHOUSE, N. B., 1970, *Q. Jl Math.*, **21**, 277 ; 1971, *Ibid.*, **22**, 277 ; 1973, *J. Phys.* A, **6**, 1115 ; 1974, *J. math. Phys.*, **15**, 119.
BACKHOUSE, N. B., and BRADLEY, C. J., 1970, *Q. Jl Math.*, **21**, 203 ; 1972, *Ibid.*, **23**, 225.
BACON, G. E., 1962, *Neutron Diffraction* (Oxford : Clarendon Press).
BARKER, A. S., and LOUDON, R., 1972, *Rev. mod. Phys.*, **44**, 18.
BARKER, A. S., and TINKHAM, M., 1962, *Phys. Rev.*, **125**, 1527.
BEGUM, N. A., CRACKNELL, A. P., JOSHUA, S. J., and REISSLAND, J. A., 1969, *J. Phys.* C, **2**, 2329.
BELL, M. I., 1972, *J. Phys.* C, **5**, L279.
BELOV, N. V., and KUNTSEVICH, T. S., 1971, *Acta crystallogr.* A, **27**, 511.
BERTAUT, E. F., 1968 a, *Acta crystallogr.* A, **24**, 217 ; 1968 b, *Helv. phys. Acta*, **41**, 683 ; 1971, *J. Phys.*, *Paris*, *Coll.* C, **1**, 462.
BERTAUT, E. F., CHAPPERT, J., MARÉSCHAL, J., REBOUILLAT, J. P., and SIVARDIÈRE, J., 1967, *Solid St. Commun.*, **5**, 293.
BERTAUT, E. F., FRUCHART, D., BOUCHAUD, J. P., and FRUCHART, R., 1968, *Solid St. Commun.*, **6**, 251.
BERTAUT, E. F., and MARÉSCHAL, J., 1967, *Solid St. Commun.*, **5**, 93.
BETHE, H. A., 1928, *Annln Phys.*, **87**, 55 ; 1929, *Ibid.*, **3**, 133.
BHAGAVANTAM, S., and VENKATARAYUDU, T., 1948, *Theory of Groups and its Application to Physical Problems* (Waltair, India : Andhra University Press ; reprinted New York : Academic Press, 1969).
BIRMAN, J. L., 1962 a, *Phys. Rev.*, **125**, 1959 ; 1962 b, *Ibid.*, **127**, 1093 ; 1963, *Ibid.*, **131**, 1489 ; 1966 a, *Ibid.*, **150**, 771 ; 1966 b, *Phys. Rev. Lett.*, **17**, 1216 ; 1967, *Chem. Phys. Lett.*, **1**, 343.
BIRMAN, J. L., LAX, M., and LOUDON, R., 1966, *Phys. Rev.*, **145**, 620.
BIRSS, R. R., 1963, *Rep. Prog. Phys.*, **26**, 707 ; 1964, *Symmetry and Magnetism* (Amsterdam : North-Holland).

BOARDMAN, A. D., O'CONNOR, D. E., and YOUNG, P. A., 1973, *Symmetry and its Applications in Science* (London : McGraw-Hill).

BOERSCH, H., GEIGER, J., and STICKEL, W., 1966, *Phys. Rev. Lett.*, **17**, 379.

BOUCKAERT, L. P., SMOLUCHOWSKI, R., and WIGNER, E., 1936, *Phys. Rev.*, **50**, 58.

BRADLEY, C. J., 1966, *J. math. Phys.*, **7**, 1145 ; 1973, *J. Phys. A*, **6**, 1843.

BRADLEY, C. J., and CRACKNELL, A. P., 1970, *J. Phys. C*, **3**, 610 ; 1972, *The Mathematical Theory of Symmetry in Solids : Representation Theory for Point Groups and Space Groups* (Oxford : Clarendon Press).

BRADLEY, C. J., and DAVIES, B. L., 1968, *Rev. mod. Phys.*, **40**, 359 ; 1970, *J. math. Phys.*, **11**, 1536.

BRINKMAN, W. F., and ELLIOTT, R. J., 1966 a, *J. appl. Phys.*, **37**, 1457 ; 1966 b, *Proc. R. Soc. A*, **294**, 343.

BROWN, E., 1964, *Phys. Rev. A*, **133**, 1038.

BUERGER, M. J., 1963, *Elementary Crystallography* (New York : Wiley).

BÜLOW, R., NEUBÜSER, J., and WONDRATSCHEK, H., 1971, *Acta crystallogr. A*, **27**, 520.

BURSTEIN, E., JOHNSON, F. A., and LOUDON, R., 1965, *Phys. Rev. A*, **139**, 1239.

CASELLA, R. C., and TREVINO, S. F., 1972, *Phys. Rev. B*, **6**, 4533.

CHEREMUSHKINA, A. V., and VASIL'EVA, R. P., 1966, *Fizika tverd. Tela*, **8**, 822. English translation : *Soviet Phys. solid St.*, **8**, 659.

CHIA, K. K., CRACKNELL, A. P., and WALKER, R. J., 1974, *J. Phys. F*, **4**, 1121.

CLAUS, R., 1970, *Physics Lett. A*, **31**, 299.

COCHRAN, W., 1960, *Adv. Phys.*, **9**, 387.

COCHRAN, W., COWLEY, R. A., DOLLING, G., and ELCOMBE, M. M., 1966, *Proc. R. Soc. A*, **293**, 433.

CORNWELL, J. F., 1966, *Phys. kondens. Materie*, **4**, 327 ; 1969, *Group Theory and Electronic Energy Bands in Solids* (Amsterdam : North-Holland) ; 1971, *Phys. Stat. Sol.*, **43**, 763 ; 1972, *Ibid.*, **52**, 275.

COWLEY, R. A., 1962, *Phys. Rev. Lett.*, **9**, 159 ; 1964, *Phys. Rev. A*, **134**, 981.

CRACKNELL, A. P., 1966 a, *Aust. J. Phys.*, **19**, 519 ; 1966 b, *Prog. theor. Phys., Kyoto*, **35**, 196 ; 1968 a, *Applied Group Theory* (Oxford : Pergamon) ; 1968 b, *Adv. Phys.*, **17**, 367 ; 1969 a, *Proc. Camb. phil. Soc. math. phys. Sci.*, **65**, 567 ; 1969 b, *J. Phys. C*, **2**, 500 ; 1969 c, *Ibid.*, **2**, 1425 ; 1969 d, *Ibid.*, **2**, 1764 ; 1969 e, *Rep. Prog. Phys.*, **32**, 633 ; 1970 a, *Phys. Rev. B*, **1**, 1261 ; 1970 b, *J. Phys. C*, **3**, *Metal Phys. Suppl.*, S 175 ; 1971 a, *The Fermi Surfaces of Metals* (London : Taylor & Francis) ; 1971 b, *J. Phys. C*, **4**, 2488 ; 1971 c, *Adv. Phys.*, **20**, 747, *erratum* **21**, 691 ; 1972, *Acta crystallogr. A*, **28**, 597 ; 1973 a, *J. Phys. C*, **6**, 826 ; 1973 b, *Ibid.*, **6**, 841 ; 1973 c, *Ibid.*, **6**, 1054 ; 1973 d, *Phys. Rev. B*, **7**, 2145 ; 1974 a, *Magnetism in Crystalline Materials* (Oxford : Pergamon) ; 1974 b, *Thin Solid Films*, **21**, 107 ; 1974 c, *J. Phys. F*, **4**, 466 ; 1974 d, *Thin Solid Films*, **24**, 279.

CRACKNELL, A. P., and CHIA, K. K., 1973, *Proc. Second Cairo Solid St. Conf.* (in press).

CRACKNELL, A. P., CRACKNELL, M. F., and DAVIES, B. L., 1970, *Phys. Stat. Sol.*, **39**, 463.

CRACKNELL, A. P., and JOSHUA, S. J., 1968, *J. Phys. A*, **1**, 40 ; 1969 a, *Proc. Camb. phil. Soc. math. phys. Sci.*, **66**, 493 ; 1969 b, *Phys. Stat. Sol.*, **36**, 737 ; 1970, *Proc. Camb. phil. Soc. math. phys. Sci.*, **67**, 647.

CRACKNELL, A. P., and SEDAGHAT, A. K., 1972, *J. Phys. C*, **5**, 977 ; 1973, *Ibid.*, **6**, 2350.

CRACKNELL, A. P., and WONG, K. C., 1967, *Aust. J. Phys.*, **20**, 173 ; 1973, *The Fermi Surface : Its Concept, Determination, and Use in the Physics of Metals* (Oxford : Clarendon Press).

DANIEL, M. R., and CRACKNELL, A. P., 1969, *Phys. Rev.*, **177**, 932.

DARWIN, C. G., 1914, *Phil. Mag.*, **27**, 675.

DAVIES, B. L., and LEWIS, D. H., 1971, *Phys. Stat. Sol. A*, **7**, 523.

DAVISSON, C. J., and GERMER, L. H., 1927, *Phys. Rev.*, **30**, 705.

DEDERICHS, P. H., 1972, *Solid St. Phys.*, **27**, 135.

DE GROOT, S. R., 1951, *Thermodynamics of Irreversible Processes* (Amsterdam : North-Holland).

DE WAMES, R. E., and VREDEVOE, L. A., 1967, *Phys. Rev. Lett.*, **18**, 853.

DIMMOCK, J. O., 1963, *J. math. Phys.*, **4**, 1307.

DIMMOCK, J. O., and WHEELER, R. G., 1962 a, *Phys. Rev.*, **127**, 391 ; 1962 b, *J. Phys. Chem. Solids*, **23**, 729 ; 1964, *The Mathematics of Physics and Chemistry*, Vol. 2, edited by H. Margenau and G. M. Murphy (New York : Van Nostrand), p. 725.

DOLLING, G., COWLEY, R. A., and WOODS, A. D. B., 1965, *Can. J. Phys.*, **43**, 1397.

DONI, E., and PASTORI PARRAVICINI, G., 1973, *J. Phys. C*, **6**, 2859.

DONNAY, G., CORLISS, L. M., DONNAY, J. D. H., ELLIOTT, N., and HASTINGS, J. M., 1958, *Phys. Rev.*, **112**, 1917.

DONNAY, J. D. H., DONNAY, G., COX, E. G., KENNARD, O., and KING, M. V., 1963, *Crystal Structure Data, Determinative Tables* (New York : American Crystallographic Association).

DUKE, C. B., and TUCKER, C. W., 1969, *Surf. Sci.*, **15**, 231.

DZIALOSHINSKIĬ, I. E., 1964, *Zh. éksp. teor. Fiz.*, **46**, 1420. English translation : *Soviet Phys. JETP*, **19**, 960.

EGELSTAFF, P. A., 1965, *Thermal neutron scattering* (New York : Academic Press).

ELLIOTT, R. J., 1971, *Proc. Second Int. Conf. Light Scattering in Solids*, Paris (Paris : Flammarian Sciences), p. 354.

ELLIOTT, R. J., HARLEY, R. T., HAYES, W., and SMITH, S. R. P., 1972, *Proc. R. Soc. A*, **328**, 217.

ELLIOTT, R. J., and LOUDON, R., 1960, *J. Phys. Chem. Solids*, **15**, 146.

ELLIOTT, R. J., SMITH, S. R. P., and YOUNG, A. P., 1971, *J. Phys. C*, **4**, L317.

ELLIOTT, R. J., and STEVENS, K. W. H., 1953, *Proc. R. Soc. A*, **218**, 553.

ELLIOTT, R. J., and THORPE, M. F., 1967, *Proc. phys. Soc.*, **91**, 903.

ENGLMAN, R., 1972, *The Jahn–Teller Effect in Molecules and Crystals* (London : Wiley).

EYRING, H., WALTER, J., and KIMBALL, G. E., 1944, *Quantum Chemistry* (New York : Wiley).

FADDEYEV, D. K., 1964, *Tables of the Principal Unitary Representations of Fedorov Groups* (Oxford : Pergamon).

FALICOV, L. M., and RUVALDS, J., 1968, *Phys. Rev.*, **172**, 498.

FIESCHI, R., 1957, *Physica, 's Grav.*, **23**, 972.

FLEURY, P. A., and LOUDON, R., 1968, *Phys. Rev.*, **166**, 514.

FOLLAND, N. O., and BASSANI, F., 1968, *J. Phys. Chem. Solids*, **29**, 281.

FREI, V., 1966, *Czech. J. Phys.*, **16**, 207.

FRENKEL, J., 1931, *Phys. Rev.*, **37**, 17.

FRIKKEE, E., 1973, *A Study of Excitations in Ferromagnetic Nickel* (Leiden : Thesis).

FUJIWARA, T., and OHTAKA, K., 1968, *J. phys. Soc. Japan*, **24**, 1326.

FUMI, F. G., 1952, *Nuovo Cim.*, **9**, 739.

GARD, P., 1973 a, *J. Phys. A*, **6**, 1807 ; 1973 b, *Ibid.*, **6**, 1829 ; 1973 c, *Ibid.*, **6**, 1837.

GOLDRICH, F. E., and BIRMAN, J. L., 1968, *Phys. Rev.*, **167**, 528.

GORZKOWSKI, W., 1964, *Phys. Stat. Sol.*, **6**, 521.

GRIFFITH, J. S., 1961, *The Theory of Transition Metal Ions* (Cambridge : University Press).

HAEFNER, K., STOUT, J. W., and BARRETT, C. S., 1966, *J. appl. Phys.*, **37**, 449.

HALPERN, O., and JOHNSON, M. H., 1939, *Phys. Rev.*, **55**, 898.

HAMERMESH, M., 1962, *Group Theory and its Application to Physical Problems* (Reading, Mass. : Addison-Wesley).

HAYAKAWA, K., NAMIKAWA, K., and MIYAKA, S., 1971, *J. phys. Soc. Japan*, **31**, 1408.

HEINE, V., 1960, *Group Theory in Quantum Mechanics* (Oxford : Pergamon Press) ; 1963, *Proc. phys. Soc.*, **81**, 300 ; 1964, *Surf. Sci.*, **2**, 1.

HENRY, N. F. M., and LONSDALE, K., 1965, *International Tables for X-ray Crystallography*, Vol. 1, *Symmetry Groups* (Birmingham : Kynoch).

190 A. P. Cracknell

HERRING, C., 1937, *Phys. Rev.*, **52**, 365 ; 1942, *J. Franklin Inst.*, **233**, 525 ; 1966, *Magnetism*, Vol. 4, edited by G. T. Rado and H. Suhl (New York : Academic Press), p. 1.

HERZFELD, C. M., and MEIJER, P. H., 1961, *Solid St. Phys.*, **12**, 1.

HOLSTEIN, T., and PRIMAKOFF, H., 1940, *Phys. Rev.*, **58**, 1098.

HORNREICH, R. M., 1972, *I.E.E.E. Trans. Magn.*, MAG-**8**, 584.

HOUSTON, W. V., 1948, *Rev. mod. Phys.*, **20**, 161.

HUBERMAN, B. A., and BURSTEIN, E., 1971, *AIP Conf. Proc. No. 5, Magnetism and Magnetic Materials*, p. 1350.

HUTCHINGS, M. T., 1964, *Solid St. Phys.*, **16**, 227.

INDENBOM, V. L., 1960, *Kristallografiya*, **5**, 513. English translation : *Soviet Phys. Cryst.*, **5**, 493.

JAHN, H. A., 1938, *Proc. R. Soc.* A, **164**, 117.

JAHN, H. A., and TELLER, E., 1937, *Proc. R. Soc.* A, **161**, 220.

JAN, J. P., 1972, *Can. J. Phys.*, **50**, 925.

JANSEN, L., and BOON, M., 1967, *Theory of Finite Groups. Applications in Physics* (Amsterdam : North-Holland).

JANSSEN, T., 1973, *Crystallographic Groups* (Amsterdam : North-Holland).

JENNINGS. P. J., 1970, *Surf. Sci.*, **20**, 18 ; 1971, *Ibid.*, **26**, 509.

JENSEN, J., 1971, *Int. J. Magn.*, **1**, 271.

JOHNSON, F. A., and LOUDON, R., 1964, *Proc. R. Soc.* A, **281**, 274.

JONES, H., 1960, *The Theory of Brillouin Zones and Electronic States in Crystals* (Amsterdam : North-Holland).

JONES, R. O., 1966, *Proc. phys. Soc.*, **89**, 443.

JOSHUA, S. J., and CRACKNELL, A. P., 1969, *J. Phys.* C, **2**, 24.

JUDD, B. R., 1963, *Operator Techniques in Atomic Spectroscopy* (New York : McGraw-Hill).

KAMBE, K., 1967, *Z. Naturf.* A, **22**, 322.

KARAVAEV, G. F., 1964, *Fizika tverd. Tela*, **6**, 3676. English translation : *Soviet Phys. solid St.*, **6**, 2943 ; 1966, *Izv. vȳssh. ucheb. Zaved., Fiz.*, (3), 64. English translation : *Soviet Phys. J.*, **9**, 37.

KATIYAR, R. S., 1970, *J. Phys.* C, **3**, 1087.

KILLINGBECK, J., 1970, *Rep. Prog. Phys.*, **33**, 533 ; 1972, *J. Phys.* C, **5**, 985.

KITTEL, C., 1963, *Quantum Theory of Solids* (New York : Wiley).

KLEINER, W. H., 1966, *Phys. Rev.*, **142**, 318 ; 1967, *Ibid.*, **153**, 726 ; 1969, *Ibid.*, **182**, 705.

KNOX, R. S., and GOLD, A., 1964, *Symmetry in the Solid State* (New York : Benjamin).

KOEHLER, W. C., WOLLAN, E. O., and WILKINSON, M. K., 1960, *Phys. Rev.*, **118**, 58.

KOPTSIK, V. A., 1966, *Shubnikov Groups. Handbook on the Symmetry and Physical Properties of Crystal Structures* (Moscow : University Press).

KOSTER, G. F., 1957, *Solid St. Phys.*, **5**, 173.

KOSTER, G. F., DIMMOCK, J. O., WHEELER, R. G., and STATZ, H., 1963, *Properties of the Thirty-two Point Groups* (Cambridge, Mass. : M.I.T. Press).

KOVALEV, O. V., 1965 a, *Irreducible Representations of the Space Groups* (New York : Gordon & Breach) ; 1965 b, *Fizika tverd. Tela*, **7**, 103. English translation : *Soviet Phys. solid St.*, **7**, 77.

KRISTOFEL', N. N., 1964, *Fizika tverd. Tela*, **6**, 3266. English translation : *Soviet Phys. solid St.*, **6**, 2613.

KUDRYAVTSEVA, N. V., 1967, *Fizika tverd. Tela*, **9**, 2364. English translation : *Soviet Phys. solid St.*, **9**, 1850 ; 1968, *Fizika tverd. Tela*, **10**, 1616. English translation : *Soviet Phys. solid St.*, **10**, 1280 ; 1969, *Fizika tverd. Tela*, **11**, 1031. English translation : *Soviet Phys. solid St.*, **11**, 840.

KUPER, C. G., and WHITFIELD, G. D., 1963, *Polarons and Excitons* (Edinburgh : Oliver & Boyd).

KUYATT, C. E., and SIMPSON, J. A., 1967, *Rev. scient. Instrum.*, **38**, 103.

LAMPRECHT, G., and MERTEN, L., 1969, *Phys. Stat. Sol.*, **35**, 353.

LANDAU, L. D., 1937 a, *Zh. éksp. teor. Fiz.*, **7**, 19; 1937 b, *Ibid.*, **7**, 627. English translations in *Collected papers of L. D. Landau*, edited by D. ter Haar (New York : Gordon & Breach, 1965).

LANDAU, L. D., and LIFSHITZ, E. M., 1958, *Statistical Physics* (London : Pergamon Press).

LAX, M., 1962, *Proc. Int. Conf. Semicond. Phys., Exeter*, p. 396 ; 1965, *Phys. Rev.*, **138**, 793.

LAX, M., and HOPFIELD, J., 1961, *Phys. Rev.*, **124**, 115.

LE CORRE, Y., 1958 a, *J. Phys., Paris*, **19**, 750 ; 1958 b, *Bull. Soc. fr. Minéral. Cristallogr.*, **81**, 120.

LEECH, J. W., and NEWMAN, D. J., 1969, *How to use Groups* (London : Methuen).

LEWIS, D. H., 1973, *J. Phys. A*, **6**, 125.

LITVIN, D., 1973, *Acta crystallogr. A*, **29**, 651.

LIU, S. H., 1972, *Phys. Rev. Lett.*, **29**, 793.

LOMONT, J. S., 1959, *Applications of Finite Groups* (New York : Academic Press).

LOUDON, R., 1968, *Adv. Phys.*, **17**, 243.

LOVESEY, S. W., 1972, *J. Phys. C*, **5**, 2769.

LUEHRMANN, A. W., 1968, *Adv. Phys.*, **17**, 1.

LYUBARSKII, G. YA., 1960, *The Application of Group Theory in Physics* (Oxford : Pergamon).

McCLURE, D. S., 1959 a, *Solid St. Phys.*, **8**, 1 ; 1959 b, *Ibid.*, **9**, 399.

McRAE, E. G., 1966, *J. chem. Phys.*, **45**, 3467 ; 1968, *Surf. Sci.*, **11**, 479.

McWEENY, R., 1963, *Symmetry. An Introduction to Group Theory and its Applications* (Oxford : Pergamon).

MARADUDIN, A. A., and VOSKO, S. H., 1968, *Rev. mod. Phys.*, **40**, 1.

MARSHALL, W., and LOVESEY, S. W., 1971, *Theory of thermal neutron scattering : The use of neutrons for the investigation of condensed matter* (Oxford : Clarendon Press).

MATTHEISS, L. F., 1964, *Phys. Rev. A*, **133**, 1399.

MEIJER, P. H. E., and BAUER, E., 1962, *Group Theory : the Application to Quantum Mechanics* (Amsterdam : North-Holland).

MILLER, S. C., and LOVE, W. F., 1967, *Tables of Irreducible Representations of Space Groups and Co-representations of Magnetic Space Groups* (Boulder, Col. : Pruett Press).

MONTGOMERY, H., 1969, *Proc. R. Soc. A*, **309**, 521.

MONTGOMERY, H., and CRACKNELL, A. P., 1973, *J. Phys. C*, **6**, 3156.

MONTGOMERY, H., and DOLLING, G., 1972, *J. phys. Chem. Solids*, **33**, 1201.

MONTGOMERY, H., and PAUL, G. L., 1971, *Proc. R. Soc. Edinb. A*, **70**, 107.

MULLIKEN, R. S., 1933, *Phys. Rev.*, **43**, 279.

NEUBÜSER, J., WONDRATSCHEK, H., and BÜLOW, R., 1971, *Acta crystallogr. A*, **27**, 517.

NOWACKI, W., EDENHARTER, A., and MATSUMOTO, L., 1967, *Crystal Data, Systematic Tables* (New York : American Crystallographic Association).

NYE, J. F., 1957, *Physical Properties of Crystals* (Oxford : Clarendon Press).

O'DELL, T. H., 1967, *The Electrodynamics of Magnetoelectric Media* (Amsterdam : North-Holland).

OLBRYCHSKI, K., and VAN HUONG, N., 1970, *Acta phys. polon. A*, **37**, 369.

ONSAGER, L., 1931 a, *Phys. Rev.*, **37**, 405 ; 1931 b, *Ibid.*, **38**, 2265.

OPECHOWSKI, W., 1940, *Physica, 's Grav.*, **7**, 552.

OPECHOWSKI, W., and GUCCIONE, R., 1965, *Magnetism*, Vol. 2A, edited by G. T. Rado and H. Suhl (New York : Academic Press), p. 105.

OVERHAUSER, A. W., 1956, *Phys. Rev.*, **101**, 1702.

PALMBERG, P. W., DE WAMES, R. E., and VREDEVOE, L. A., 1968, *Phys. Rev. Lett.*, **21**, 682.

PALMBERG, P. W., DE WAMES, R. E., VREDEVOE, L. A., and WOLFRAM, T., 1969, *J. appl. Phys.*, **40**, 1158.

PEREL, J., BATTERMAN, B. W., and BLOUNT, E. I., 1968, *Phys. Rev.*, **166**, 616.

PIPPARD, A. B., 1957, *The Elements of Classical Thermodynamics* (Cambridge : University Press).

POSLEDOVICH, M., WINTER, F. X., BORSTEL, G., and CLAUS, R., 1973, *Phys. Stat. Sol. B*, **55**, 711.

POULET, H., 1965, *J. Phys., Paris*, **26**, 684.

RAINFORD, B., HOUMAN, J. G., and GUGGENHEIM, H. J., 1972, *Neutron Inelastic Scattering*. Proceedings of a symposium, Grenoble (Vienna : I.A.E.A.), p. 655.

RIEDER, K. H., and HÖRL, E. M., 1968, *Phys. Rev. Lett.*, **20**, 209.

ROSENTHAL, J. E., and MURPHY, G. M., 1936, *Rev. mod. Phys.*, **8**, 317.

ROTH, W. L., 1960, *J. appl. Phys.*, **31**, 2000.

RUDRA, P., 1965, *J. math. Phys.*, **6**, 1273.

RUPPIN, R., and ENGLMAN, R., 1970, *Rep. Prog. Phys.*, **33**, 149.

SAKURAI, J., BUYERS, W. J. L., COWLEY, R. A., and DOLLING, G., 1968, *Phys. Rev.*, **167**, 510.

SANDERCOCK, J. R., PALMER, S. B., ELLIOTT, R. J., HAYES, W., SMITH, S. R. P., and YOUNG, A. P., 1972, *J. Phys. C*, **5**, 3126.

SEDAGHAT, A. K., and CRACKNELL, A. P., 1974, *Physica, 's Grav.*, **71**, 615.

SEITZ, F., 1936, *Ann. Math.*, **37**, 17.

SHERWOOD, P. M. A., 1972, *Vibrational Spectroscopy of Solids* (Cambridge : University Press).

SHOCKLEY, W., 1937, *Phys. Rev.*, **52**, 866.

SHUBNIKOV, A. V., and BELOV, N. V., 1964, *Colored Symmetry* (Oxford : Pergamon).

SIROTIN, YU. I., and MIKHEL'SON, L. M., 1968, *Fizika tverd. Tela*, **10**, 1843. English translation : *Soviet Phys. solid St.*, **10**, 1450.

SLATER, J. C., 1963, *Quantum Theory of Molecules and Solids*, Vol. 1. *Electron Structure of Molecules* (New York : McGraw-Hill) ; 1965, *Quantum Theory of Molecules and Solids*, Vol. 2. *Symmetry and Energy Bands in Crystals* (New York : McGraw-Hill ; reprinted New York : Dover, 1972).

SLATER, J. C., and KOSTER, G. F., 1954, *Phys. Rev.*, **94**, 1498.

SMITH, C. S., 1958, *Solid St. Phys.*, **6**, 175.

SÓLYOM, J., 1966, *Physica, 's Grav.*, **32**, 1243 ; 1971, *J. Phys., Paris*, **32**, Suppl. C1, 471.

SPITZER, W. G., MILLER, R. C., and KLEINMAN, D. A., 1962, *Bull. Am. phys. Soc.*, **7**, 280.

STERN, R. M., and TAUB, H., 1970, *Crit. Rev. solid St. Sci.*, **1**, 221.

STEVENS, K. W. H., 1952, *Proc. phys. Soc. A*, **65**, 209.

STREITWOLF, H. W., 1969 a, *Phys. Stat. Sol.*, **33**, 217 ; 1969 b, *Ibid.*, **33**, 225.

SUBRAMANIAN, R., 1973, *J. Phys. C*, **6**, L47.

SUGANO, S., TANABE, Y., and KAMIMURA, H., 1970, *Multiplets of Transition-metal Ions in Crystals* (New York : Academic Press).

SUZUKI, T., HIROTA, N., TANAKA, H., and WATANABE, H., 1971, *J. phys. Soc. Japan*, **30**, 888.

TINKHAM, M., 1964, *Group Theory and Quantum Mechanics* (New York : McGraw-Hill).

THOMSON, G. P., and REID, A., 1927, *Nature, Lond.*, **119**, 890.

TOKUNAGA, M., and MATSUBARA, T., 1966, *Prog. theor. Phys., Kyoto*, **35**, 581.

VAN HOVE, L., 1953, *Phys. Rev.*, **89**, 1189.

VON NEUMANN, J., 1961, 1963, *John von Neumann : Collected Works*, edited by A. H. Taub (Oxford : Pergamon Press) ; Vol. I, 1961 ; Vol. II, 1961 ; Vol. III, 1961 ; Vol. IV, 1962 ; Vol. V, 1963 ; Vol. VI, 1963.

WANNIER, G., 1937, *Phys. Rev.*, **52**, 191.

WANNIER, G. H., and FREDKIN, D. R., 1962, *Phys. Rev.*, **125**, 1910.

WARREN, J. L., 1968, *Rev. mod. Phys.*, **40**, 38.

WEYL, H., 1931, *The Theory of Groups and Quantum Mechanics* (London : Methuen).

WIGNER, E. P., 1959, *Group Theory and its Application to the Quantum Mechanics of Atomic Spectra* (New York : Academic Press).

WILSON, E. B., DECIUS, J. C., and CROSS, P. C., 1955, *Molecular Vibrations* (New York : McGraw-Hill).

WINTER, F. X., and CLAUS, R., 1972, *Optics Communications*, **6**, 22.

WONDRATSCHEK, H., BÜLOW, R., and NEUBÜSER, J., 1971, *Acta crystallogr.* A, **27**, 523.

WONDRATSCHEK, H., and NEUBÜSER, J., 1967, *Acta crystallogr.*, **23**, 349.

WOOSTER, W. A., 1973, *Tensors and Group Theory for the Physical Properties of Crystals* (Oxford : Clarendon Press).

WORLTON, T. G., and WARREN, J. L., 1972, *Computer Physics Communications*, **3**, 88.

WYCKOFF, R. W. G., 1963, *Crystal Structures*, Vol. 1 (New York : Interscience) ; 1964, Vol. 2, *Inorganic Compounds* RX_n, R_nMX_2, R_nMX_3 ; 1965, Vol. 3, *Inorganic Compounds* $R_x(MX_4)_y$, $R_x(MX_p)_y$, *Hydrates and Ammoniates* ; 1966, Vol. 5, *The Structures of Aliphatic Compounds* ; 1968, Vol. 4, *Miscellaneous Inorganic Compounds, Silicates and Basic Structural Information.*

YU, M. L., and CHANG, J. T. H., 1970, *J. Phys. Chem. Solids*, **31**, 1997.

ZAK, J., 1964 a, *Phys. Rev.* A, **134**, 1602 ; 1964 b, *Ibid.*, A, **134**, 1607 ; 1966, *Ibid.*, **151**, 464.

ZAK, J., CASHER, A., GLÜCK, M., and GUR, Y., 1969, *The Irreducible Representations of Space Groups* (New York : Benjamin).

Author Index

Abragam, A. and Bleaney, B., 65
Aizu, K., 170, 171, 172
Akhiezer, A. I., Bar'yakhtar, V. G., and
Peletminskii, S. V., 125
Altmann, S. L., 50
Altmann, S. L. and Bradley, C. J., 37, 50,
75, 93, 95, 101
Altmann, S. L. and Cracknell, A. P., 37, 50,
93, 101
Anderson, P. W. and Blount, E. I., 182
Ascher, E., 179, 180, 184, 185
Ascher, E. and Janner, A., 180
Ashby, N. and Miller, S. C., 85, 88
Astrov, D. N., 44

Backhouse, N. B., 76, 85, 186
Backhouse, N. B. and Bradley, C. J., 85, 87
Bacon, G. E., 4
Barker, A. S. and Loudon, R., 131
Barker, A. S. and Tinkham, M., 167
Barrett, C. S.—see Haefner, K.
Bar'yakhtar, V. G.—see Akhiezer, A. I.
Bassani, F.—see Folland, N. O.
Batterman, B. W.—see Perel, J.
Bauer, E.—see Meijer, P. H. E.
Begum, N. A., Cracknell, A. P., Joshua,
S. J. and Reissland, J. A., 115
Bell, M. I., 15
Belov, N. V.—see Shubnikov, A. V.
Belov, N. V. and Kuntsevich, T. S., 6
Bertaut, E. F., 160, 162, 164, 165, 166
Bertaut, E. F., Chappert, J., Maréschal, J.,
Rebouillat, J. P. and Sivardière, J., 162
Bertaut, E. F. and Maréschal, J., 162
Bethe, H. A., 2, 26, 27, 48, 153
Bhagavantam, S. and Venkatarayudu, T.,
2, 103
Birman, J. L., 137, 138, 140, 141, 177, 178,
183, 184—see also Goldrich, F. E.
Birman, J. L., Lax, M., and Loudon, R., 141
Birss, R. R., 2, 8, 11, 38, 41, 42, 43, 45
Bleaney, B.—see Abragam, A.
Blount, E. I.—see Anderson, P. W., Perel, J.
Boardman, A. D., O'Connor, D. E. and
Young, P. A., 2
Boersch, H., Geiger, J. and Stickel, W., 151
Boon, M.—see Jansen, L.
Borstel, G.—see Posledovich, M.
Bouckaert, L. P., Smoluchowski, R. and
Wigner, E., 2, 16, 30, 32, 73, 74, 93, 94,
105, 125, 147, 169
Bradley, C. J., 85, 136, 137—see also
Altmann, S. L., Backhouse, N. B.
Bradley, C. J. and Cracknell, A. P., 2, 3, 6, 8,
13, 14, 15, 16, 17, 19, 25, 32, 47, 50, 51,
53, 55, 60, 66, 69, 74, 75, 76, 77, 92, 97,
112, 136, 150

Bradley, C. J. and Davies, B. L., 8, 13, 14,
25, 77, 138
Brinkman, W. F. and Elliott, R. J., 7
Brown, E., 85, 86, 87
Buerger, M. J., 158
Bülow, R.—see Neubüser, J., Wondratschek,
H.
Bülow, R., Neubüser, J. and Wondratschek,
H., 6
Burstein, E.—see Huberman, B. A.
Burstein, E., Johnson, F. A. and Loudon,
R., 141
Buyers, W. J. L.—see Sakurai, J.

Casella, R. C. and Trevino, S. F., 146
Casher, A.—see Zak, J.
Chang, J. T. H.—see Yu, M. L.
Chappert, J.—see Bertaut, E. F.
Cheremushkina, A. V. and Vasil'eva,
R. P., 46
Chia, K. K.—see Cracknell, A. P.
Chia, K. K., Cracknell, A. P. and Walker,
R. J., 101
Claus, R., 132—see also Nitsch, W.,
Posledovich, M., Winter, F. X.
Cochran, W., 167, 169
Cochran, W., Cowley, R. A., Dolling, G.,
and Elcombe, M. M., 107
Coles, B. R., 109
Cornwell, J. F., 2, 15, 37, 75, 141
Cowley, R. A., 102, 107, 167, 168, 169—see
also Cochran, W., Dolling, G., Sakurai, J.
Cox, E. G.—see Donnay, J. D. H.
Cracknell, A. P., 2, 3, 8, 13, 14, 16, 19, 45,
46, 47, 55, 69, 73, 79, 85, 94, 96, 98,
100–1, 103, 111, 112, 116, 117, 118, 119,
121, 122, 125, 128, 129, 130, 144, 159, 170,
172, 175, 180, 185, 186—see also Altmann,
S. L., Begum, N. A., Bradley, C. J.,
Chia, K. K., Daniel, M. R., Joshua,
S. J., Montgomery, H., Sedaghat, A. K.
Cracknell, A. P. and Chia, K. K., 100
Cracknell, A. P., Cracknell, M. F., and
Davies, B. L., 185
Cracknell, A. P. and Joshua, S. J., 25,
55, 56, 112
Cracknell, A. P. and Sedaghat, A. K., 26,
146, 149, 152, 186
Cracknell, A. P. and Wong, K. C., 69, 90
Cracknell, M. F.—see Cracknell, A. P.
Cross, P. C.—see Wilson, E. B.
Curie, P., 43

Daniel, M. R. and Cracknell, A. P., 112
Darwin, C. G., 155
Davies, B. L.—see Bradley, C. J., Cracknell,
A. P.

Subject Index